Illusions of Seeing

Thomas Ditzinger

Illusions of Seeing

Exploring the World of Visual Perception

Springer

Thomas Ditzinger
Reichartshausen, Germany

ISBN 978-3-030-63634-0 ISBN 978-3-030-63635-7 (eBook)
https://doi.org/10.1007/978-3-030-63635-7

This Springer imprint is published by the registered company Springer Nature Switzerland AG
The registered company address is: Gewerbestrasse 11, 6330 Cham, Switzerland

For Leonie, the star of my eye

Foreword

Thomas Ditzinger takes us on a fascinating tour through the visual world of illusions. Light is the stimulus for the eye, but the complex processes that follow stimulation are illuminated by the study of illusions. The book not only presents a great variety of visual illusions (often in color), but it also links them to the physiological process in the visual system and to their subtle use by artists. Illusions can make us smile and wonder at the sophistication of seeing, and Ditzinger does not lose contact with the fun of the phenomena that he so clearly displays.

Psychology Department
University of Dundee
Dundee, Scotland, UK

Nick Wade

Preface

We live in a fantastic, fascinating, wonderful world. Each one of us is an important piece of the puzzle in this world and tries to get along in it as well as possible. To do this we, use our senses, which enable us to simultaneously perceive an enormous amount of environmental stimuli and information. The human system of perception has developed in the course of evolution in close interaction with its environement. Therefore, it is no wonder that our brain and its perceptual apparatus became an image of its environment—with similar fantastic, fascinating, wonderful characteristics. You can see this for yourself in this book.

In order to accomplish these enormous perceptual achievements, a highly complex self-organized configuration of the brain is required. Our brain consists of the unimaginable number of at least 100,000,000,000 nerve cells, the neurons. That is almost 20 times as much as the current number of the world's population! Together with the glial cells, whose exact function has not yet been fully clarified, the neurons form the basic building blocks of our human brain. On average, each neuron has about 1000–10,000 connections to other nerve cells. The connection runs through strongly branched axons with a total number of at least 100–1000 trillions in American naming or billions in British naming (a 1 with 14–15 zeros)—which is already in the range of the world population of ants on earth. If one added the length of all these branches of communication, it is estimated that the total distance would be an almost unbelievable half to one million kilometers. That would correspond to a telephone cable about 25 times the distance from the earth to the moon and back again, wound up in our brain!

Vision is perhaps the most important innovation of evolution for the development of the human brain and perception. Approximately 60% of all environmental experiences reach our brain through the eye. That is why vision is also regarded as the key to understanding our brain. The main aim of this book is to get to know and appreciate the wonderful abilities of our visual apparatus. It will become clear that our visual and perceptual apparatus is wonderfully capable of presenting the most contradictory and complicated environmental impressions very simply in concise forms. Let yourself be taken on a journey of discovery through your own perception! Because of the special destination, you do not only read in a book, but you read—and see—mainly in yourself. This book serves as your guide on this fantastic adventure. Similar to a holiday trip, there will be eyes of relaxation, but also highly impressive natural beauties and interesting discoveries to experience.

Again and again it will become clear that our perception can be characterized by a few simple basic characteristics: It is pragmatic and always strives for the simplest solution (one could also say it is "lazy"), and it has a powerful wealth of prior knowledge and prejudices learned from evolution. You will see that the most impressive illusions of seeing always take place when these basic characteristics come into conflict with each other. You will recognize the important role of a third characteristic of your perception: It is imaginative and meaningful and accepts compromises—and sometimes cheats a little.

This adventure guide for the eyes presents to you a multitude of well-known and new optical illusions, illusions, and fantastic images. Each of these travel experiences will provide you with a new way to understand your own perception.

In the first journey you will learn something about the nature of light and its perception by the human eye. You can also get to know sheep that not only bleat and eat grass, but also know the laws of vision. You will learn some important fundamentals of Gestalt psychology and the essential laws of human perception.

The second journey takes you into the magic world of geometric optical illusions. You will be amazed at how easy it is to lead your brain up the garden path with simple line drawings. Thus, straight lines suddenly appear tilted, curved, or of different lengths.

The third journey is about the perception of forms, brightness, and transparency, and their interaction. You will realize that the perception of form and brightness of a figure depends decisively on the form and brightness of the background. For example, you can see two identical suns, which are perceived completely differently, and experience what "brightness contrast enhancement" is all about.

The fourth journey is dedicated to ambiguous perception. After a short viewing time, you see images that suddenly look completely different than before and come to life. Cubes that jump like wild in space, or drawings of young people that look just like their own grandparents, will fascinate you. These ambivalent images are used, among other things, to measure very individual "fingerprints of your mind."

The fifth journey into color vision tells you why all cats are grey at night, why the sky is blue, why day vision differs from night vision and what complementary colors, and afterimages are all about. And you can meet a dog whose picture waggles its tail in front of your eyes. Also read why men are much more color-blind than women, and why the color blue is something very special.

During the sixth journey into spatial vision you can learn why you have two eyes. You can penetrate into deeper perceptual dimensions and experience various methods of depth perception such as the Pulfrich effect, the red-green anaglyph technique, pairs of random dot images, and autostereograms.

The seventh journey involves seeing of movement and its interaction with color, form, and spatial depth. We transform a solid pencil into rubber and black and white images into color and set stationary images in motion depending on their color.

The eighth journey takes you into the everyday life of our completely normal life. We are in the supermarket, the dentist, the fashion house, looking for well-camouflaged hidden animals, in shadow and light. You will be amazed how many small and big miracles and illusions even the greyest everyday life consists of, if you only look with open eyes. And sometimes our daily life also brings leisure and sports or a holiday. Therefore, we also visit a football stadium, a ski slope, San Francisco, and Loch Ness.

The upcoming trips will take you from the polar star to Jupiter via Freiburg back to the moon and the sea, to the green meadow of the local honeybee, to Venice Beach and to a nearby motorway bridge.

You can travel all these distances with the most fascinating, cheapest, and most comfortable known means of transport: Your own thoughts. The entire route ahead of you, from the infinite expanses of the universe to the café around the corner, fits exactly into this book and into your head—just like the rolled out "telephone cable" of the aforementioned nerve connections.

I promise you a lot of fun and things to marvel on all these paths!

Reichartshausen, Germany Thomas Ditzinger

Contents

Chapter 1
First Journey: Light, Perception, and the Laws of Seeing

In the beginning there was light. "God said, let there be light. And there was light. God saw that the light was good. God separated the light from the darkness" (Genesis 1:3–4).

Light—wonder of life, sign of enlightenment, symbol of meaning and purpose. The way of light is the way to life.

From a physical point of view, light is electromagnetic radiation that is 900,000 times faster than sound in empty space until it meets an obstacle. However, humans are only able to perceive a tiny fraction of this electromagnetic spectrum.

In this first journey through the world of perception, you will become acquainted with the physical and psychological laws of seeing that determine our experience and are responsible for how we see what we see.

1.1 Human Perception

Man's astonishing success in the evolution of species is mainly due to his extraordinary ability to adapt quickly and well to a wide variety of environmental conditions. For example, he is able to absorb information sent out by the environment through his senses and thus create in his brain as natural an image as possible of his outside world. This fascinating process is called *perception*.

Perception makes use of the most diverse sensory organs. As you read these lines, perhaps the smell of the finished dinner is pouring out of the oven into your nose, perhaps you are hearing a dog barking in the neighborhood, perhaps you are feeling some warming rays of sunlight on your skin, perhaps you are palpating the back of this book. A corresponding perception can cause the brain to react if necessary—for example, to the perception of something burnt in the oven.

We live in a land of milk and honey full of information and news. Our environment sends out stimuli of all kinds in abundance. We take up this diversity of stimuli

T. Ditzinger, *Illusions of Seeing*, https://doi.org/10.1007/978-3-030-63635-7_1

with our sensory organs, in the development of which nature has shown a great inventiveness. Nature has exploited the most diverse physical possibilities in the most intelligent way and has invented hearing, smelling, tasting, touching, feeling, sense of balance, sensation of temperature and pain, internal organ perception, and perception by the immune system. But the latest and most successful invention of evolution is seeing.

1.2 Light and Vision

Light is the carrier of visual information. Compared to other carrier media of human perception, such as odorous substances or sound waves, light has invaluable advantages. This includes above all its tremendous speed: For example, light is about 900,000 times faster than sound. While we always perceive sound with a certain delay, we can perceive all events taking place in the field of vision practically simultaneously with our eyes. This brings with it a huge early warning advantage in dangerous situations. In addition, light is not as susceptible to interference as sound, which is very difficult to perceive when there is a headwind or loud background noise, for example.

We are surrounded by a sea of light; light is omnipresent. Its significance for man can already be deduced from the history of creation as depicted in the Old Testament:

"God said, let there be light. And there was light. God saw that the light was good. God separated the light from the darkness" (Gen. 1:3–4). But what exactly is it about the light?

Although all of us believe we know what light is, it is not easy to describe this natural phenomenon even in the age of modern science. Right into the eighteenth century, science was "even in the dark about light"—including Benjamin Franklin, who wrote this quote.

Light is always emitted from a natural or artificial transmitter. Examples of natural light sources are the sun, stars, fire, and chemiluminescence in fireflies; examples of artificial sources are light bulbs, fluorescent tubes, and candlelight. All these transmitters generate energy, which they emit into their environment in the form of *electromagnetic waves*.

The exact process of the emission of light can be explained by an interplay between vibrating charged matter in the transmitter, electric, and magnetic fields. The faster the matter oscillates in the transmitter, the higher the radiated energy and the faster the electromagnetic waves oscillate. The faster a wave oscillates, the more and the shorter wave packets it brings. It is also clear that these waves have shorter *wavelengths*.

The energy of the electromagnetic wave is thus directly related to the wavelength: the smaller the wavelength, the greater the energy of the wave.

The wavelength determines the perception of the color of the light. We see light with a low wavelength of about 400 nm (0.00004 cm) as blue light, while we perceive light with a slightly higher wavelength of about 800 nm as red.

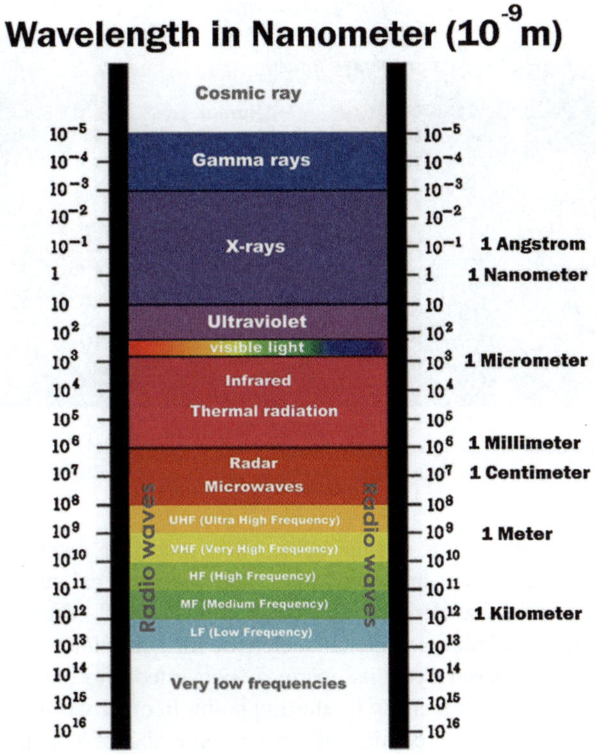

Fig. 1.1 The electromagnetic spectrum

Normally, the light emitted by a source has a whole spectrum of different wavelengths in different intensities or brightness. Any radiation with wavelengths outside this narrow range is invisible to humans! The visible light is therefore only a small part of the entire electromagnetic spectrum, which is shown in Fig. 1.1.

Radiation with less energy than the visible red is, for example, infrared radiation, microwaves, and radio waves. Radiation with higher energy than the visible blue is, for example, ultraviolet, gamma radiation, and cosmic radiation.

Electromagnetic radiation has the property of maintaining itself and its energy in a vacuum without the aid of matter in the form of a wave. This wave propagates from the source with the very high speed of light, constant for the vacuum, of approximately 300,000 km/s. The speed of light is constant for the vacuum. Like all electromagnetic radiation, light moves in one direction until it hits an obstacle, and the radiation transports energy from one place to another.

This is comparable to an elevator or a taxi transporting people from one place to another.

An energy exchange between the light wave and matter takes place at the obstacle. Depending on the internal condition of the obstacle, the radiation is absorbed by

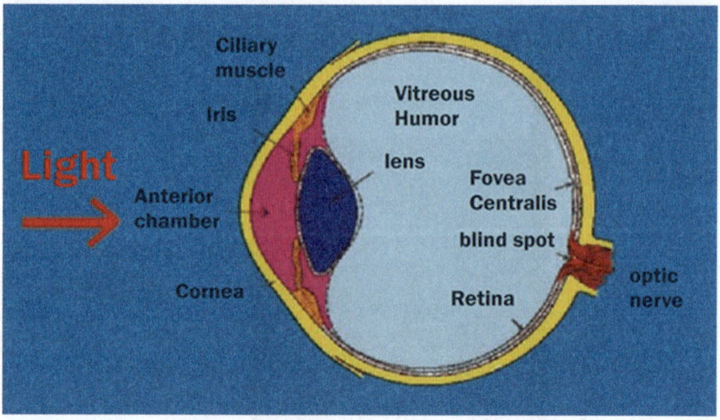

Fig. 1.2 A schematic cross-section through the eye

the material with very specific wavelengths. This process of absorbing electromagnetic radiation by substances and converting their energy is called *absorption*.

All rays with other wavelengths are reflected unchanged. The wavelengths and intensity of this reflected radiation are characteristic for a substance - like a fingerprint. When such a beam hits our eyes, its energy is converted into chemical energy and nerve impulses, and we can identify its starting point. In other words: We can *see* it.

How does this sensory perception of seeing come about in detail? To achieve a visual impression, the human eye uses a variety of techniques. With the help of a number of fantastic features, it has found a way to pick up and transmit incoming rays with a highly intelligent information processing system.

As can be seen in Fig. 1.2, the visual process begins when a light beam enters the optical apparatus of the eye. This consists of the cornea, the anterior chamber of the eye, and the lens. The light beam is refracted in the cornea and the lens. In front of the lens is the circular *iris*, also known as rainbow skin because of its striking color. The iris has a hole in its center, the *pupil*; through this, the light enters the lens. The pupil shrinks in intense light and expands in low light. Thus, the iris fulfills exactly the function of an aperture.

The shape of the lens and thus its refractive power can be changed with the help of the *ciliary muscle* surrounding it in a ring. The ciliary muscle is one of the most active muscles of our body. By tensioning the ciliary muscle, the lens becomes round, which increases its refractive power. This focuses the eye at close distances. Conversely, when the ciliary muscle relaxes, the lens is flattened and the visual acuity is adjusted to a distance. As the lens becomes more and more inelastic under constant strain, the ciliary muscle is no longer able to focus the lens at close distances (this happens naturally with increasing age). The lens images the ray of light after it has passed through the transparent *vitreous body* onto the *retina*. The retina lies on the back inner surface of the eye. There an upside-down, reduced picture is created.

The retina is equipped with many light-sensitive photoreceptors that can convert the received light into electrical nerve impulses. In total there are about 126 million (!) photoreceptors on the retina! A distinction can be made between the *rods* and the *cones*, which consist of different light-sensitive materials and which have been prepared for the experience of different brightness and colors.

The approximately 120 million rods of the retina are specialized in weak light. However, the rods cannot distinguish colors, so that in weak twilight only different shades of grey are recognized ("at night all cats are grey").

The rods are evenly distributed over the retina—apart from the *Fovea centralis*, in which no rods can be found. The fovea is a tiny pit located in the central area of the retina. There, in a very small space, are a large part of the approximately 6 million cones with which we see during the day. The cones enable us to see in color and to see sharp.

The cones and rods transmit their visual information to the neurons in the form of electrical impulses. In these neurons, which are arranged in several layers, lie *horizontal cells*, *bipolar cells*, *amacrins*, and *ganglion cells*, which assume the highly complicated task of pre-processing the image information.

Remarkably, the photoreceptors are located at the "wrong" end of the retina, i.e., at the end that is away from the light. As can be seen in Fig. 1.3, the incident light must therefore first pass through the entire processing (transparent) layers of the retina until it hits the cones and rods.

This position of the photoreceptors has two advantages for the eye. On the one hand, the layer of photoreceptors is protected against shocks and deformations. On the other hand, the photoreceptors can easily be supplied by the *pigment epithelial*

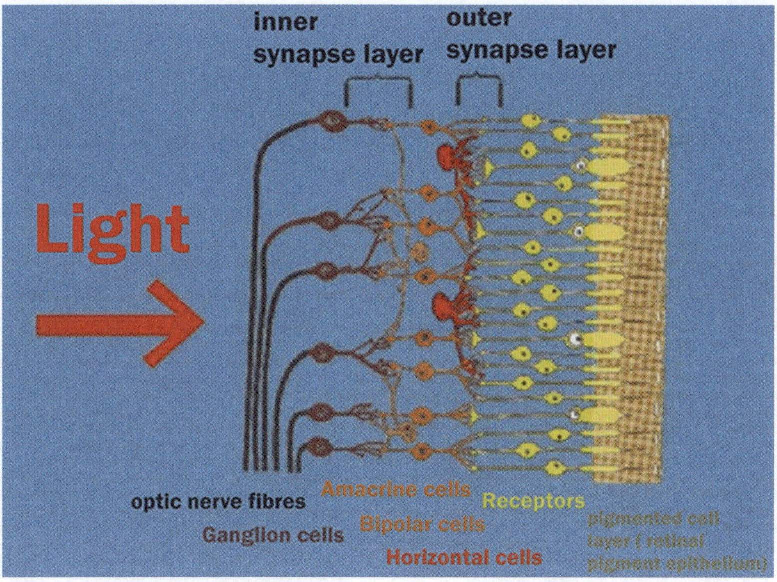

Fig. 1.3 Cross-section through the retina

cells from the side facing away from the light. The complicated visual process requires a constant reconstruction of the color pigments in the photoreceptor cells. This is done with the aid of an enzyme—a protein—contained in these pigment epithelial cells.

The retina already takes over a large part of the image evaluation itself by combining the information units of the 126 million photoreceptors in the form of electrical stimuli in the optic tract, which consists of "only" about 800,000 nerve lines. This enormous performance requires almost unbelievable capabilities of the downstream retinal layers.

It is no coincidence that the retina in the embryonic phase develops from a protrusion of the brain tissue, i.e., it is part of the brain. The results of the visual process of an eye are the electrical impulses bundled in the 800,000 nerve fibers. They are transmitted in the optical strand to the visual center in the brain.

The optical strand passes through the retina at a certain point, the *blind spot*. At this point no photoreceptor can sit and therefore no visual perception can take place. This blind spot can be seen in Fig. 1.4.

Please close your left eye and look with your right eye at the bus driver in Fig. 1.4! Slowly move the book back and forth until the bus—which is ready for our beginning adventure trip through our perception—disappears! If it is still partially visible, rotate the image a little around the center and change the viewing distance. Also pay attention to the horizontal pattern of the garage in the background—the pattern is complemented throughout by our perception system. The other way around, the trick works the same way. Keep your right eye closed and look at the coach! This time you can make the bus driver disappear with the correct book distance!

How does our perception manage to compensate so well for the lack caused by the blind spot that we don't notice it at all in everyday life? The simplest explanation would be that one eye compensates for the blind spot of the other eye. However, the fact that the blind spot is so well balanced by the perceptual system even in one-eyed vision, making it not normally noticeable even then, speaks against this.

Fig. 1.4 Let either the bus or the bus driver disappear

Taking into account all visual information around the blind spot, our visual system completes the gap in such a way that a "good shape" is created in the simplest possible way. For Fig. 1.4, this means that the bars of the garage at the point where the coach has disappeared are continuously supplemented and no gaps in the field of vision are visible!

This fantastic quality is a first indication of the ingenious strategies of perception used by our brain, the traces of which we want to follow in the course of this journey of a somewhat different kind: the striving for "good form" and order.

1.3 The Senses Strive for Order

Now take a seat in your mind on one of the seats of the coach shown in Fig. 1.4. The journey of discovery through the fantastic land of the laws of seeing can begin!

1.3.1 A Puzzle Animal

The journey is soon stopped by a structure consisting of black and white blobs, which stands on the street in front of the bus. The view through the windowpane onto this structure is shown in Fig. 1.5.

The bus driver stops and looks anxiously at the clock. The travelers should already actually be on the meadow with his favorite animals, the sheep. But the driver knows that the sheep are patient and keep their peace, which is why they are his favorite animals. From his experience with earlier travel groups, he knows that this annoying puzzle animal, shown in Fig. 1.5, only disappears from the path when

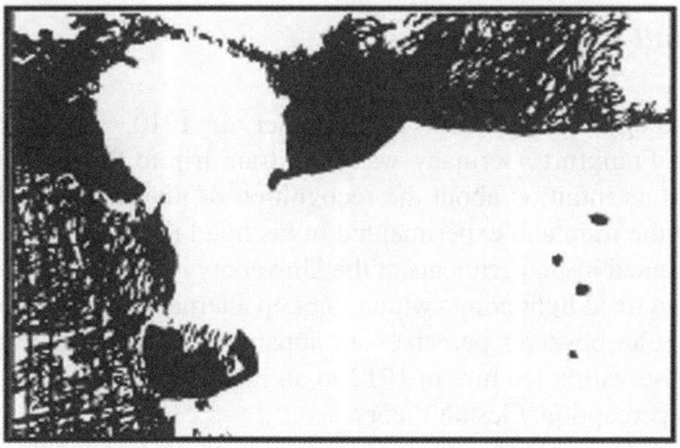

Fig. 1.5 The puzzle animal on the street (Dallenbach's figure)

all the travelers have *not only seen it*, but also *perceived* it! In order to speed up this process, he explains the following to his passengers:

> Each of you can recognize this animal, if you only look long enough at the initially arbitrary combination of black and white areas. Give your perceptual system enough time; it will then automatically find a meaningful interpretation of this confusion!

If you follow this advice, after some time (between a few seconds and hours!!) the presumed confusion disappears suddenly and your perception changes into a state of order.

The puzzle animal transforms into the picture in Fig. 1.11, trots comfortably to the side with a loud "moo," and clears the way for the bus.

On the basis of this incomplete puzzle picture, we can exactly understand the success strategy of our perception. When you look at the picture for the first time, you will surely "only" see a complete mess of black and white areas. However, our perceptual system is very reluctant to be satisfied with this state of supposed disorder. Rather, it is constantly in search of an ordered sensation. This "inner restlessness" is suddenly replaced by the perception of a meaningful interpretation of the individual surfaces as an overall picture. This condition proves to be permanently stable.

Once you have perceived this ordered structure, if you look at the picture again after some time, you will have no more problems recognizing it!

This characteristic of supplementing disordered, incomplete or—as in the blind spot case—completely missing sensory impressions into a meaningful overall impression is part of a large, fantastic plan that our perceptual system pursues.

The best way to describe this plan is to strive for a meaningful perception that is as simple as possible. In the words of Wolfgang Metzger, this state of order can also be described as the "favorite child of the senses." Metzger is a representative of *Gestalt psychology*, which has set itself the task of interpreting this plan of our system of perception. This will be discussed below.

1.3.2 Gestalt Psychology

Gestalt psychology was born in the summer of 1910, when Professor Max Wertheimer of Frankfurt, Germany was on a train trip to the Rhineland. All of a sudden, he had an intuition about the recognition of movements and sham movements. He left the train and experimented in his hotel room with some kind of flip book. He continued his experiments at the University of Frankfurt. He investigated the effect of two fixed light points which light up alternately in rapid succession. He recognized that an observer perceives an apparent movement between these two points. This observation led him in 1912 to an important insight into the organization of human perception: Gestalt theory.

Gestalt theory means that in the stable state of perception we do not perceive a sum or sequence of individual sensations, but the image as a whole—with the individual sensations as their components. This view corresponds exactly to the processes when looking at the Dallenbach figure (Fig. 1.5): The whole (the *figure*) is more than the sum of its parts!

For the first time, Wertheimer was able to explain the perception of a figure with the help of *grouping laws*. Gestalt psychologists such as Kurt Koffka, Wolfgang Köhler, Wolfgang Metzger, and Michael Stadler continued this work and recognized that the system of perception summarizes the individual image components according to such grouping laws into forms.

To demonstrate these grouping laws of seeing, the fantastic coach on its journey to the land of the laws of seeing now stops at a pasture with many sheep.

1.4 The Sheep and the Laws of Sight

"Today you're almost on time," the shepherd greets the bus driver, and his dog wags his tail. The dog is happy because he knows that he has something to do very right away.

"My sheep are very special sheep. They don't just eat the whole long day, stand all over the meadow, bleat stupidly and appear like ordinary sheep in people's dreams. No, my sheep are very eager to demonstrate the laws of sight to you," says the shepherd.

1.4.1 The Law of Conciseness

The dog now drives the sheep, and quickly they stand—seen from far above—in an arrangement as in Fig. 1.6 on the left in front of us.

Probably you can clearly see a square and an ellipse on the left. To illustrate these forms, the dog drives the sheep apart, as can be seen in the middle. But why do we recognize these figures and not any others, as for example in Fig. 1.6 on the right? The answer is summarized in the central law of Gestalt theory, the *law of conciseness*.

The law of conciseness was formulated by Kurt Koffka (1886–1941) in a similar way:

The psychological organization will always be as good as the prevailing conditions allow. The word *good* includes characteristics such as regularity, symmetry, unity, uniformity, equilibrium, maximum simplicity, scarcity, and the tendency to orient vertically or horizontally.

Fig. 1.6 The law of conciseness: Surely the arrangement on the left is more like an ellipse and a square (as you can see in the center) than the shapes you can see on the right

Fig. 1.7 Do you recognize two superimposed triangles instead of a complicated dodecagon?

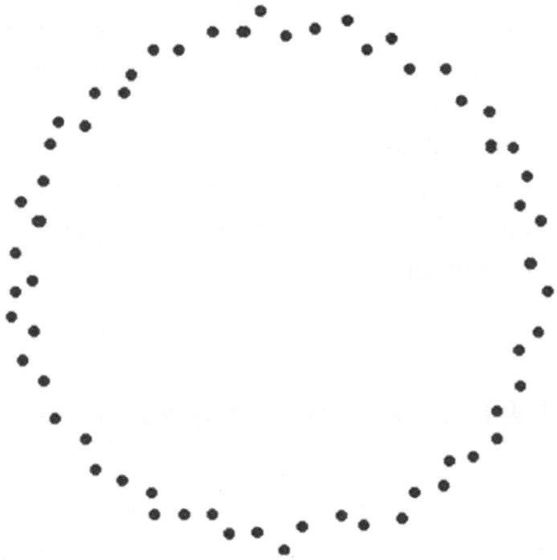

Fig. 1.8 A few points that deviate from a circle line are still perceived as belonging to the circle line!

The conciseness law is very general, which ensures its central role as the supreme rule for the perception of figures. It is also called the *law of good form* or *the law of simplicity*. The result of viewing any scene is always such that the final structure perceived is as simple as possible.

Therefore, in the first arrangement of the sheep we clearly recognize the ellipse and the square. These two patterns are characterized by their simplicity compared to all other possible patterns. The same goes for us when we look at Fig. 1.7, in which we again look at the flock of sheep from a great distance from above. Surely you will very soon recognize two simple triangles lying on top of each other instead of a complicated dodecagonal structure!

Experience has shown that circles, right angles, and straight lines are very good forms in the sense of conciseness. Examples of the preferential perception of these good forms are shown by the sheep in the next orders. Seen from far above they appear as in Figs. 1.8, 1.9 and 1.10.

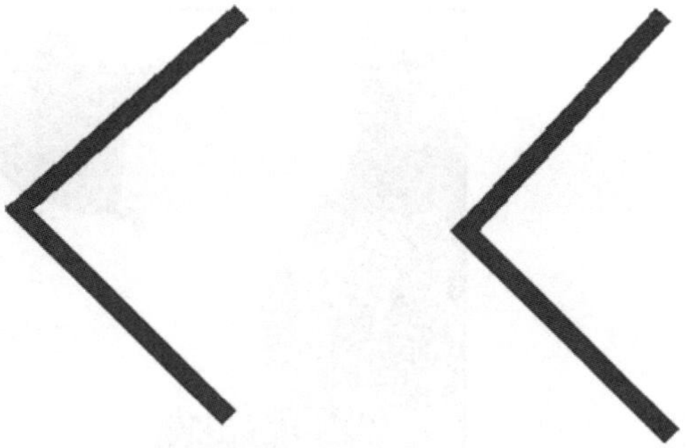

Fig. 1.9 An angle of 87 degrees or 93 degrees is still recognized as a right angle

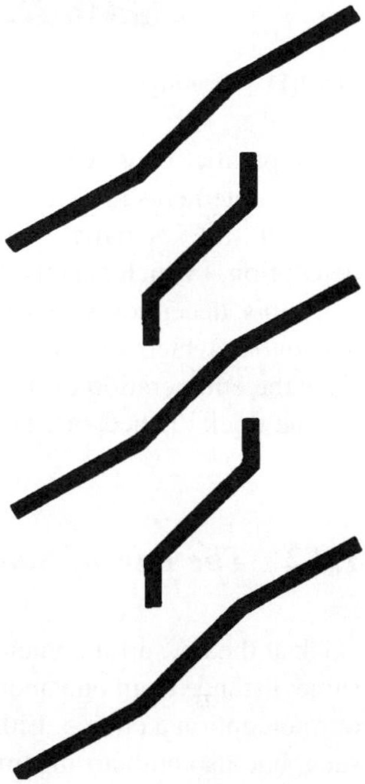

Fig. 1.10 Lipp's deception: The lines appear tilted against each other

It can be seen, for example (Fig. 1.8), that a few points that deviate slightly from a circular line appear as if they were really lying on the circle—this is, of course, not the case. You can also see (Fig. 1.9) that angles of 87° or 93° are still perceived as forming a right angle (90°).

To demonstrate the good shape of straight lines, the sheep have come up with something special. Look at the result in Fig. 1.10.

According to its discoverer, this pattern is called *Lipp's deception*. The principle of conciseness has the effect that our perception bridges bent lines and perceives them as straight lines. Lipp's deception proves that this is indeed the case. The central segments of the individual lines consist mainly of long straight pieces, which are perceived as strongly tilted against each other. In reality, however, they lie exactly parallel to each other! This effect is provided by the short, kinked end pieces

Fig. 1.11 The resolution of Dallenbach's figure (cf. Fig. 1.5)

of the parallel lines, which alternately point upwards and downwards. Our perception complements these bent lines to a straight line as best we can. This is only possible if the actually parallel middle lines are tilted accordingly in the perception—which exactly causes the deception!

Lipp's deception is an example of numerous fascinating *geometrical-optical deceptions* which we will turn to in our second journey. But first we want to complete the enumeration of the laws of seeing.

The flock of sheep are now showing us these laws one by one.

1.4.2 The Law of Similarity

Look at the new arrangements of the lambs in Fig. 1.12. The lambs have exactly the same distance from one another vertically and horizontally. This arrangement gives our perception a choice. Either horizontal or vertical arrangements of sheep can be seen, but also both arrangements at the same time!

This option disappears immediately if every second column consists of black sheep (cf. part of Fig. 1.12). Now you immediately perceive a clear vertical arrangement of black and white sheep!

This perception clarifies the *law of similarity*. It states that the individual elements of an image are preferably perceived as a group if they are similar. This similarity may relate to color, brightness, size, orientation, or shape. The same grouping effect would be achieved by alternating large and small sheep or red and green sheep.

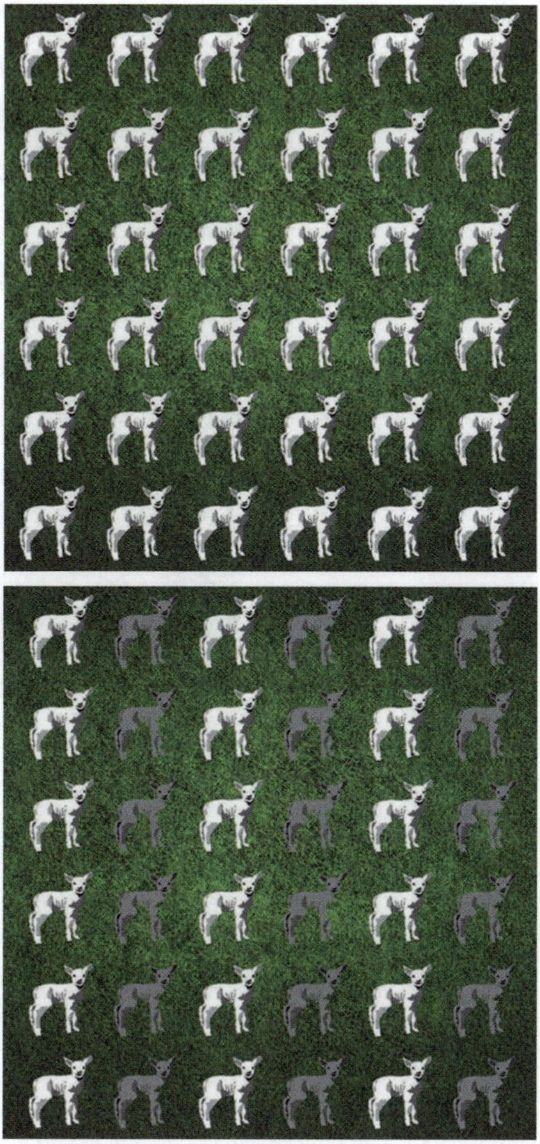

Fig. 1.12 The law of similarity. *Above*: Horizontal and vertical arrangements of sheep are perceived equally. *Below*: Surely you will immediately recognize clearly vertical arrangements of black and white sheep

1.4.3 The Law of Proximity

When some sheep are closer to each other than others, the structure immediately catches the eye. If the vertical distance (upper part of Fig. 1.13) is very low, you will surely recognize an alignment of the sheep in vertical columns. This important structuring principle is called *the law of proximity*. In other words, stimuli or picture elements that are close together are easily perceived as belonging together.

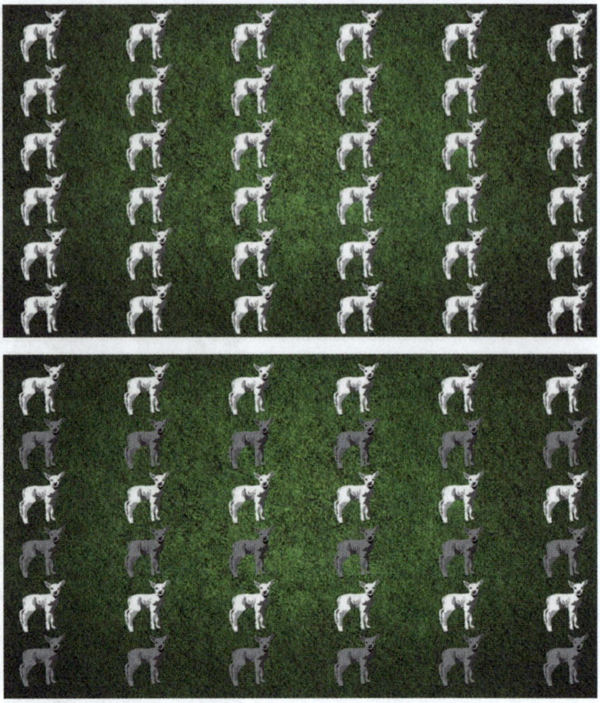

Fig. 1.13 The law of proximity

A look at the sheep in Fig. 1.13 below shows how powerful this grouping principle is, in which the law of proximity overrides the law of similarity.

1.4.4 The Law of Good Continuation

Again the sheep have adopted a new arrangement, which you can see in Fig. 1.14 from a bird's eye view.

On the left, you can clearly see two intersecting curves, which extend from A to B and from C to D. But it could just as well be two lines from A to C and from B to D. The lines, however, show a bend, which obviously brings them a clear disadvantage for their perception. This perceptual property is described by *the law of good continuation*. This law states that lines with a continuous straight line or a line with as little curvature as possible are best grouped into a unit.

Figure 1.14 in the middle shows that the assumption of a continuous curve is firmly anchored in our perception. Surely you are assuming that the rectangle

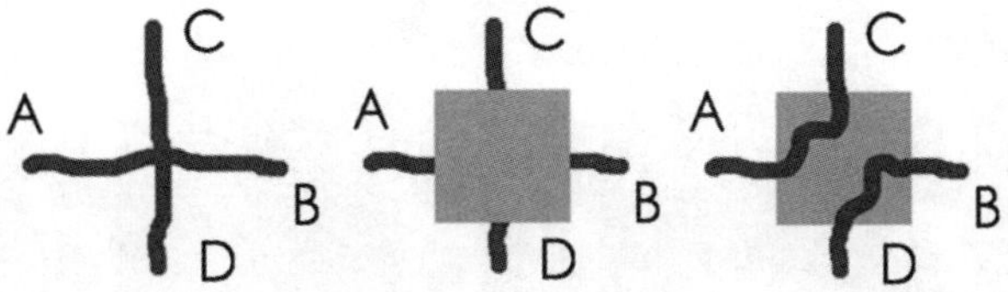

Fig. 1.14 The law of good continuation explains the clear perception of the orders

covers two straight lines that lead from A to B and from C to D. Figure 1.14 on the right shows that this is not necessarily the case.

1.4.5 The Law of Unity

Figure 1.15 above shows a pattern consisting of two adjacent vertical groups of sheep. This perception changes immediately when, as in Fig. 1.15 below, the columns with longer distance are closed. Now clearly closed formations are recognized. Closed lines unite more easily to form a shape than those that do not enclose a surface. This statement is the content of the *law of unity*. Its strength is evident in Fig. 1.15 below: Here the law of unity overrides the law of proximity.

1.4.6 The Law of Experience

Prior knowledge and experience play an important role in the grouping of picture elements.

So an illiterate person would certainly not be able to group the lines in Fig. 1.16 in the same way and complete the missing lines as you who can read them. Such a perceptual achievement describes the *law of experience*.

Also, our bus driver has his experiences. And these say that he must go on to introduce his guests to the next adventures. The sheep are now also very exhausted and glad that they no longer must demonstrate the remaining *law of common destiny*. This is the common perception of elements that are moved in the same direction. This demonstration would have required some additional movement from them. So, they just say goodbye to the departing coach and then turn back into a normal flock of sheep. They eat grass again, stand all over the meadow, bleat, and appear in our dreams. But we don't have long to dream, because we are on an adventure trip.

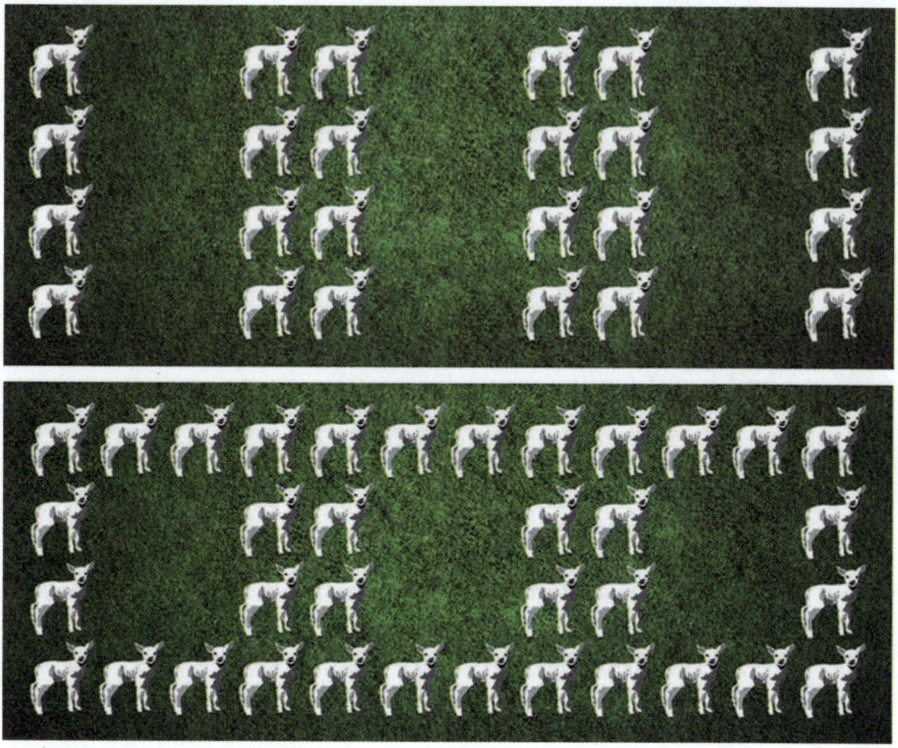

Fig. 1.15 The law of closedness

Fig. 1.16 The law of experience

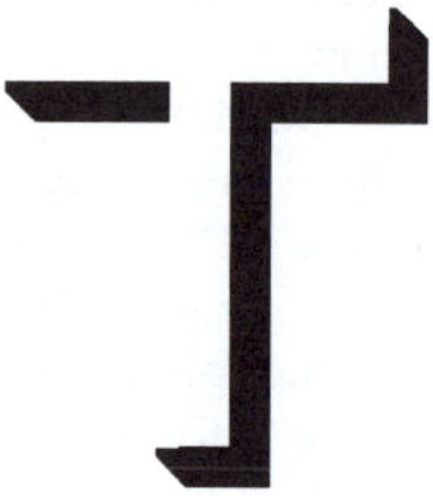

Chapter 2
Second Journey: The Geometrical-Optical Illusions

The right perspective is crucial! Depending on the perspective from which we see things, we interpret them in this or that way.

Whenever our senses play a trick on us and the spontaneous perception of reality seems to overstretch our imagination, we become overly aware of how susceptible our brain is to disruption and how easily it can be influenced. How often do we naively encounter supposed realities that turn out to be clumsy illusions after a while!

This journey leads us to the geometrical-optical illusions, which were mainly discovered at the end of the eighteenth century, but which still today disseminate a very special fascination to the observer. Most of the pictures are simple line drawings and thus form a basic framework for the analytical understanding of all further illusions of seeing.

2.1 Errors of the Senses

The optical illusions lead our senses astray. Images of optical illusions are sometimes perceived by the viewer completely differently than the actual physical conditions of the image would suggest. For example, two identical geometric figures may appear differently sized or long; straight lines may appear tilted or as if they had an interruption. As you will see, nothing can be relied upon anymore!

The best-known optical illusions are the geometrical-optical illusions. Most of the more than 200 known figures that evoke geometrical-optical illusions were discovered at the end of the eighteenth century and named after their discoverer. The way in which they break the laws of seeing—as well as the final visual result—varies from picture to picture. Hardly any other field of perceptual psychology is as ambiguous and characterized by contradictory attempts at explanation as the geometrical-optical illusions. With some pictures you don't know at all how and

T. Ditzinger, *Illusions of Seeing*, https://doi.org/10.1007/978-3-030-63635-7_2

why a certain deception occurs. This may be strange on the one hand, but on the other hand it has something calming about it. For, as is well known, ignorance is the mother of all adventures, and it is those tracks we are following with this book.

The geometrical-optical illusions normally consist of two figurative components, mostly line drawings. One part causes the deception, and the other part is the rest of the image about which one is deceived.

For example, take a look at the famous *Müller-Lyer Deception* in Fig. 2.1.

2.2 The Müller-Lyer Deception, Part 1

The Müller-Lyer illusion consists of two identical lines that are evaluated completely differently by our perception. The trigger for this is the different termination of the lines at their ends. The upper line in Fig. 2.1 is terminated by lines at acute angles, while the lower line is terminated by blunt angles. Surely the upper line seems to be a lot shorter than the lower one!

How can such a deception be explained?

As with most other geometrical-optical illusions, different explanations are also possible here. Perhaps the simplest explanation is that our perceptual system is inherently designed to assign a three-dimensional interpretation to all sensory impressions. Due to the different terminations, a different spatial depth is assigned to the two edges. The upper edge is perceived to be closer to the observer than the lower edge.

Why is that? As we already know, the orderliness of our senses strives to recognize forms that are as concise as possible. Such an ordered state is undoubtedly present at right angles! Therefore, our perceptual system tries to reinterpret arbitrary angles—if somehow possible—into right angles in a perspective view. At acute angles, such as in the upper part of Müller-Lyer's deception, this leads to the horizontal line appearing to lie in front of the paper plane. The acute angles in flat vision appear in spatial vision as perspective right angles tilted backwards. For the same reason, the lower horizontal line is perceived as lying further back. The previously

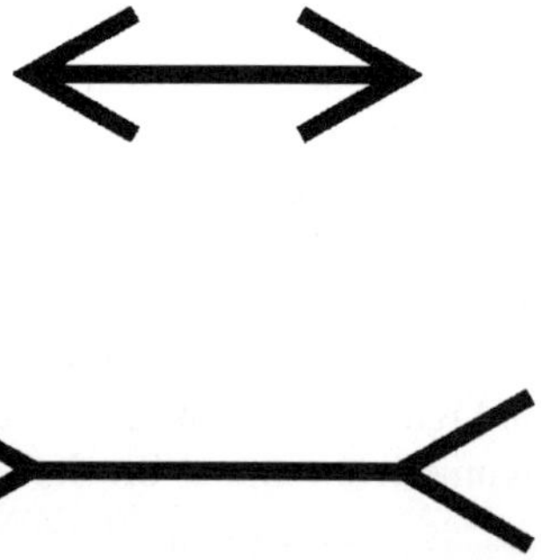

Fig. 2.1 The Müller-Lyer deception (1889): The two horizontal lines are exactly the same length!

flat angles can now be perceived as forward-protruding perspective right angles. The so-called *size constancy* finally results in the different length impression.

2.3 The Constancy of Size: An Important Basis for Perception

The phenomenon of size constancy plays an outstanding role in perception and we will encounter it again and again in the forthcoming chapters of our travel book. The constancy of size is a very special "service" of our perception system to the brain. Only with this ability is it possible to make exact size estimates and find our way in the three-dimensional world.

Because of their various distances and orientations, we recognize the size of all things in our environment as different than their actual size. With the principle of size constancy, our perceptual system pursues the ambitious goal of balancing these spatial influences. Things are corrected by our perception in their size to their real dimensions! This principle can be observed very well in the *Ponzo illusion* in Fig. 2.2.

The Ponzo illusion can be explained by the spatial order of magnitude. The two oblique lines in Fig. 2.2 on the left give the strong spatial impression of two parallel lines which have their vanishing point at great depth, as is the case with railway tracks (as shown in Fig. 2.2 on the right). Therefore, the upper crossbeam seems to be at a great spatial depth compared to the lower one. The principle of size constancy ensures that the upper bar appears much larger to us. The constancy of size enables us to perceive identical sensory impressions as different in size depending on their apparent depth—such as the two identical vertical lines in Müller-Lyer's deception: if we see the line very close, we perceive it smaller than if we assign the same line to a greater depth.

Fig. 2.2 The Ponzo illusion: the upper crossbar appears much longer than the lower one of the same length!

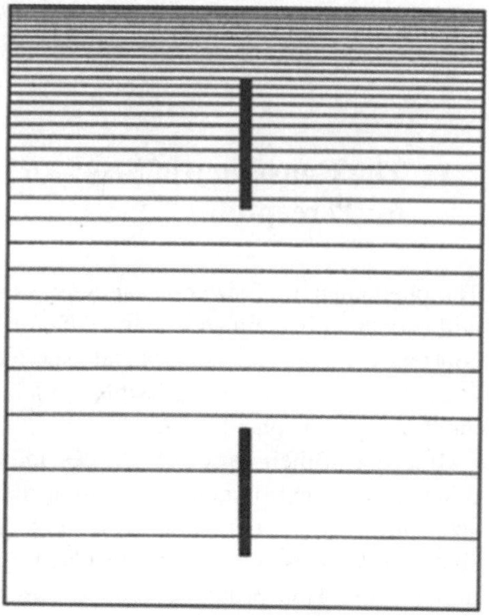

Fig. 2.3 The Gillam illusion: the upper vertical line appears much longer than the identical lower!

A further example for size constancy is the *Gillam deception* (Gillam 1986) in Fig. 2.3. The upper vertical line appears strongly extended in comparison to the equally long lower line!

Through a clever trick, a spatial scene is created here—namely through a constant increase in the density of the horizontal lines. The observer involuntarily assumes that the strokes have a constant spatial distance—the reason for this, by the way, is again the principle of conciseness that you have already come to know. The image, which is actually flat, is forced to be converted into a spatial background! Against this background, a vertical line is classified as the larger one the further down it seems to be. The perspective length of a vertical line corresponds exactly to the number of cross lines it crosses. Through the mechanism of size constancy, our perception system corrects the actual size of the line falling on the retina to a size determined by the spatial impression.

2.4 The Müller-Lyer Deception, Part 2

The spatial explanation of Müller-Lyer's deception becomes obvious in Fig. 2.4. The Müller-Lyer illusion is interpreted as a building corner, once from the outside and once from the inside. Surely the vertical center bar on the right photo appears longer than the one on the left photo. In reality, however, the two beams are exactly the same length. The explanation for this deception lies in the spatial approach.

In addition to the spatial explanation strategy, Müller-Lyer's deception, like most other geometrical-optical deceptions, can also be explained via the neuronal

Fig. 2.4 The explanation of the Müller-Lyer deception from a spatial point of view

mechanisms in the visual center. To prove this, we get to the bottom of the deception with an experiment. The images are deliberately altered or broken down into their individual components in order to analyze their effect on the viewer. This "experimental analysis" of illusions can provide us with direct clues to the mechanisms in the visual center. Figure 2.5 contains such gradually changed Müller-Lyer figures.

The length of the vertical lines appears smaller and smaller from left to right. The following conclusions can be drawn immediately from the different combinations of the components of deception used:

- Blunt angles seem to lengthen the line, but acute angles reduce it.
- The more extreme the angles, the more extreme their influence on the deception.
- The influence of blunt angles on deception is stronger than the influence of acute angles. Therefore, the second vertical line from the left appears slightly longer than the "neutral" line to its right with the vertical end lines.

These findings, gained by simple means, allow far-reaching conclusions to be drawn about the connections of information processing in the perceptual system. To do that, you have to know that neighboring photoreceptors of the retina are interconnected in a very sophisticated way in so-called detector cells. These detector cells can, for example, detect lines of a very specific gradient. When two lines meet, these

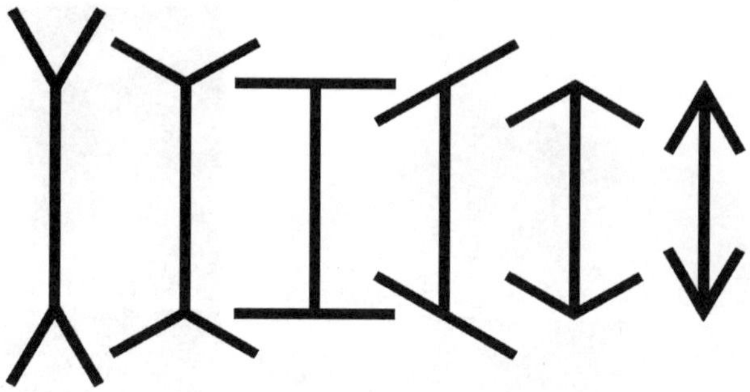

Fig. 2.5 The "experimental" investigation of Müller-Lyer's deception: from left to right, the deceptive elements are specifically altered. The length of the vertical lines appears smaller and smaller!

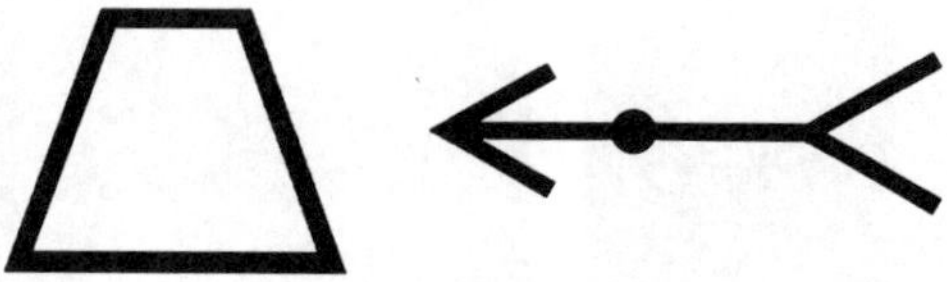

Fig. 2.6 The influence of blunt and acute angles. *Left*: Blunt angles lengthen a line; acute angles shorten it. *Right*: Judd deception. The point in the middle seems to be strongly shifted in the direction of the acute angle!

detectors interact with each other. Depending on the angular size, the detectors inhibit or amplify each other so that the lines involved appear in different sizes. Our experiments on Müller-Lyer's deception allow us to draw the following conclusion:

Acute angles inhibit the detector cells, but blunt angles stimulate them. And the detector cells manipulated in this way are ultimately responsible for the length impression of the observed line. Thus, parts of a line that end at an acute angle appear shortened. And line parts that end at an obtuse angle appear extended!

The influence of angles on the length perception of lines can be observed in various images. Two examples can be seen in Fig. 2.6. In a flat view, a trapezium can be seen on the left, consisting of a long and a short base with acute and obtuse angles. What happens now when the principle of conciseness ensures that spatial perception is carried out?

Surely you will soon see a rectangle whose upper edge seems to lie in the background. The lower line, on the other hand, seems to be farther forward. From the different angles in this perspective, right angles arise.

The influence of angles on the length of horizontal lines is also clearly visible. The acute angles shorten the lower line, just as the obtuse angles lengthen the upper line.

Exactly this effect also explains the *Judd deception*, which can be seen in Fig. 2.6 on the right. The central point in the middle of the line is considered to be strongly shifted in the direction of the acute angles! Again, the blunt angles "lengthen" the line, while the acute ones "shorten" it. Therefore, in our perception, the center slips in the direction of the acute angles!

But don't be deceived by these general insights gained from Müller-Lyer's figure! The psychology of perception, like physiology, is miles away from an unambiguous analysis of geometrical-optical illusions.

Already for the complete description of the Müller-Lyer's deception, the knowledge gained so far is far from sufficient. For example, research is already puzzled by the observed different weighting of the influence of obtuse and acute angles on the perception of length. In the Müller-Lyer's figure, for example, the obtuse angles contribute two to three times as much to the creation of the illusion as the acute angles!

By looking at only slightly modified arrangements (Figs. 2.7 and 2.8) we already reach the limits of our knowledge about Müller-Lyer's deception.

Figure 2.7 shows a *three-dimensional* Müller-Lyer deception. The three books depicted stand on the floor at the same depth and have an identical distance. Nevertheless, the left interspace looks greatly shortened and the right one enlarged, just as in the two-dimensional Müller-Lyer deception. The middle book appears shifted clearly to the left. Of course, this effect can no longer be explained by a different depth allocation of the books and the resulting misinterpreted constancy of size! In addition, the coupling of the detector cells of different edges can no longer be the cause for the deception because there are no solid edges!

One possible explanation would be that our perceptual system is able to "think" about nonexistent edges and then treat them exactly the same as real ones. Such illusory contours will be encountered in the third journey.

We encounter a similar explanatory distress in Fig. 2.8, which shows a Müller-Lyer deception with circular line ends.

As you can imagine, the two horizontal lines here are, again, the same length!

Nevertheless, a striking difference in the perception of size can again be seen here!

Fig. 2.7 The three-dimensional Müller-Lyer deception: The left book distance appears much shorter than the right book distance of the same size

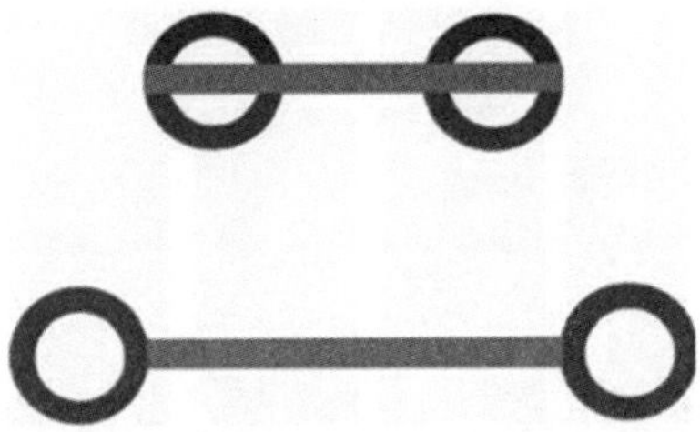

Fig. 2.8 Delboeuf's deception (1892): a Müller-Lyer deception with circular line end. Again, the two lines are the exact same length!

From the consideration of the presented modifications of the Müller-Lyer's deception we can draw the conclusion that a complete theory for the psychological explanation of all optical deceptions is probably far away. The deceptive images have far too different and diverse forms for this, and the different influencing factors in the form of competing grouping laws of seeing are far too complex.

Although the laws of seeing serve experimental psychology of perception as a tool for the isolated observation of specific characteristics, to try to explain the illusions caused by a picture in its entirety might still be a hopeless endeavor.

Here again the basic knowledge of Gestalt psychology is confirmed: The whole picture represents something completely different than the mere sum of its individual parts.

It is probably just this lack of uniformity of the geometric-optical illusions that is the main reason for the playful fascination it effects on the viewer.

2.5 The Poggendorff Deception

Figure 2.9 shows the famous *Poggendorff's illusion* as another example of a geometrical-optical illusion. It consists of a vertical bar that crosses a *diagonal continuous* line from bottom left to top right.

Continuous? Surely you don't see a continuous line, but two lines interrupted at different heights. Nevertheless, it is a continuous line. You can easily convince yourself of this with the help of a ruler.

Due to the two vertical lines, the inclined line at the bottom left seems to lie in a different spatial plane than the one at the top right. This becomes clear in Fig. 2.10, where a possible spatial perception is illustrated by auxiliary lines. The two oblique lines appear offset from each other in the room both in height and depth.

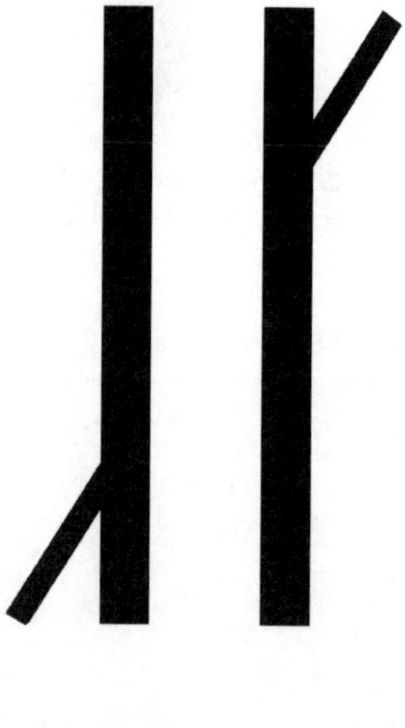

Fig. 2.9 Poggendorff's deception (1860): Is the oblique line continuous?

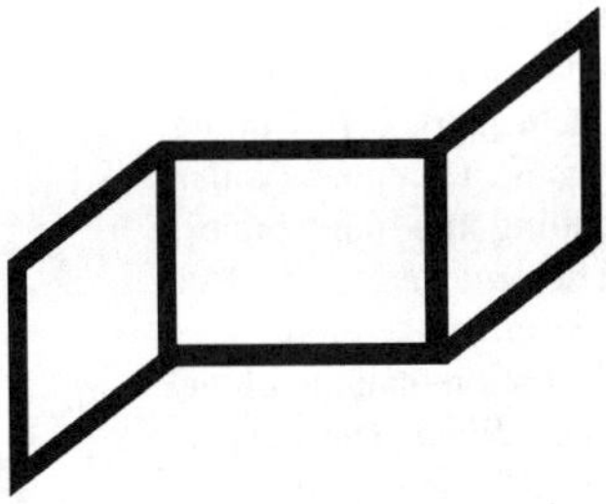

Fig. 2.10 The explanation of Poggendorff's deception by a spatial interpretation: The two oblique lines are not perceived continuously, but staggered in depth

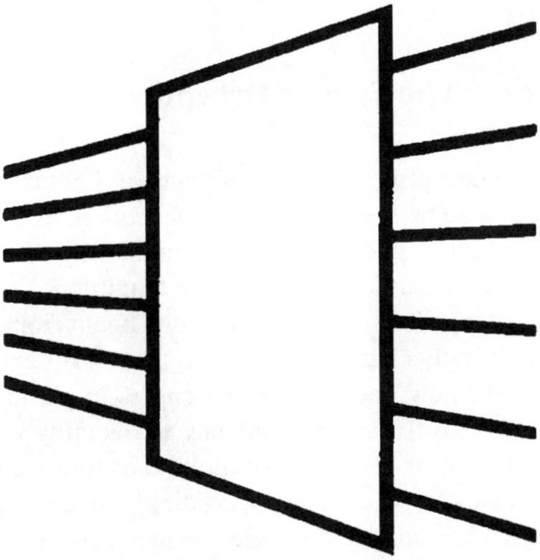

Fig. 2.11 Modified Poggendorff's deception: The deception disappears almost completely due to the clear spatial specification: only continuous diagonal lines are perceived!

If the image template of Poggendorff's figure is changed as in Fig. 2.11, the figure is perceived as spatial from the outset. The whole structure extends from the back left to the front right into the depth; only continuous diagonal lines are perceived.

Poggendorff's deception can be experimentally broken down into individual parts. For example, the influence of the acute and obtuse angles can be considered separately (Fig. 2.12 on the left and in the middle). It becomes clear that the deception is again mainly caused by the obtuse angles, which considerably "lengthen" the lines involved. This creates the impression that the intersections of the oblique lines are offset in height. The acute angles, on the other hand, "shorten" the lines involved again. As can be seen in Fig. 2.12 on the left, they cause an opposite deception. The continuation of the diagonal line at the top right seems to be even below the expected height!

Another characteristic, which we already noticed in Müller-Lyer's figure, we can perceive in Fig. 2.12 on the right. If the slope of the oblique line is increased, the extent of the deception is also increased!

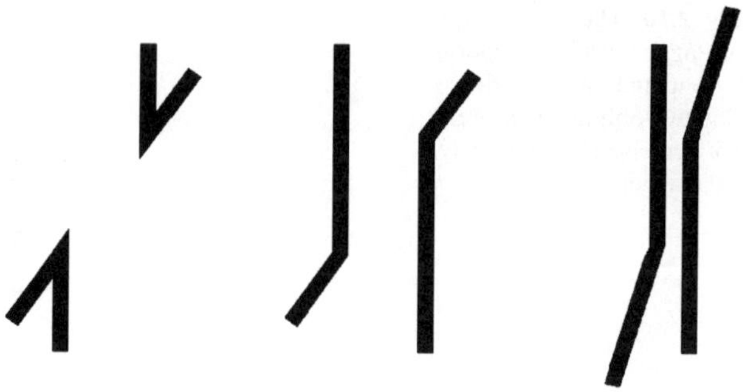

Fig. 2.12 The Poggendorff deception is taken apart. *Left*: Cutting into acute angles. *Center*: cutting into blunt angles. The acute angles have far less influence on the deception than the blunt ones! *Right*: The steeper the oblique line, the stronger the illusion!

2.6 The Sander Deception

The deceptive influence of acute and obtuse angles on the perception of a distance can also be seen in *Sander's deception*. The diagonal in the large parallelogram on Fig. 2.13 appears much longer than that in the small parallelogram. But the two diagonals are exactly the same length!

A possible explanation results again from the property of the angles involved to seemingly change the line lengths (cf. for example Fig. 2.5).

We now know that blunt angles in particular play a decisive role in deception. They have a strong tendency to seemingly enlarge the lines involved. Since the diagonal of the large parallelogram forms an obtuse angle at both intersections, it appears to be stretched accordingly. In contrast, the diagonal of the small parallelogram has only acute angles, which leads to its apparent shortening!

By the way, this approach also explains a general phenomenon described by Metzger (Metzger 2009): in every oblique parallelogram the longer diagonal is shortened and the shorter one is extended!

2.7 Hering's Deception

Look at *Hering's deception* in Fig. 2.14 on the left. Deceptions of this type are mostly very impressive specimens, which in turn—as with Sander's deception—are based on the interaction of intersecting lines. However, here not only a single line is considered, but a whole bundle of lines. This bundle causes an apparent curvature of lines crossing the bundle.

In Fig. 2.14 top left, the two lines intersecting the beam appear curved from the center; however, the two lines are exactly parallel! You can easily convince yourself of this if you look at the image from the side, with the book just below eye level, in the direction of the two straight lines. The same happens with the square above the line bundle (Fig. 2.14, top right). The straight edges of the square appear clearly "inflated" from the center!

Fig. 2.13 Sander's deception (1926): The diagonal in the large parallelogram seems significantly longer than the diagonal in the small one. In fact, they are the same length

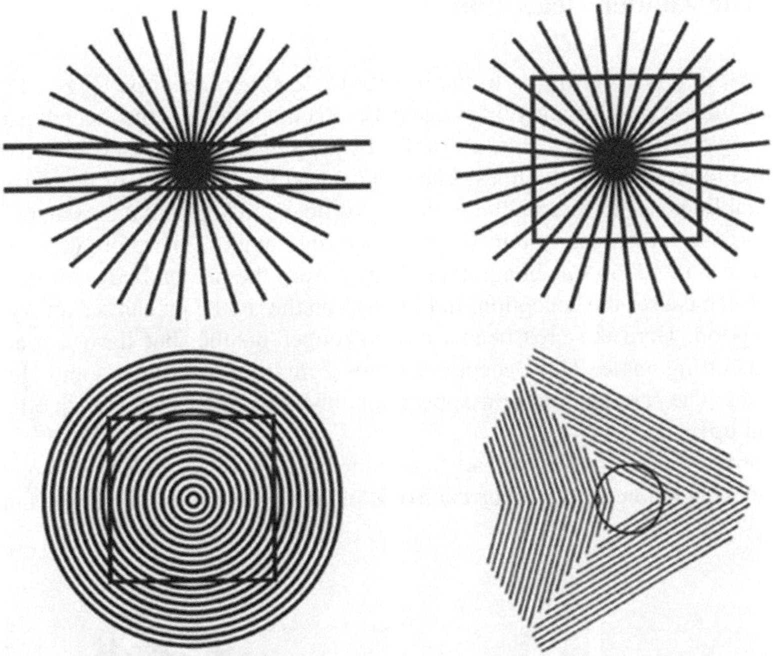

Fig. 2.14 Apparent line curvature caused by a bundle of beams. *Top left*: Hering's deception (1861): The two transverse lines appear as if they were bent away from the beam center. In reality they are parallel and straight! You can easily convince yourself of this by keeping the book flat and looking from the right along the straight line. *Top right*: The square above the line bundle appears inflated when viewed from the beam center. *Bottom left*: The reverse effect occurs when a square lies over a composition of concentric circles: The edges of the square are apparently bent inward from the center. *Bottom right:* drastic effect of the line curvature; the ideal circle appears strongly deformed

The opposite effect can be observed in Fig. 2.14, bottom left: A square is depicted above a collection of concentric circles. Surely you can immediately see the difference: The edges of the square now appear to be bent inward! Figure 2.14 bottom right shows a particularly drastic consequence of the line curvature: you can see a strongly deformed potato—which in reality is an ideal circle!

These deceptions can again be explained by the striving of our senses for a concise form—you will surely remember the law of conciseness.

Our perceptual system "longs" for right angles of intersections. It fulfils this wish as well as possible by changing the sense of orientation and tilting the lines accordingly. Since this tilting process takes place simultaneously at all intersections, the overall impression of a curved line is created!

2.8 The Zöllner Deception

Our perception is very similar to the impressive *Zöllner deception* (Fig. 2.15 left). The long lines appear strongly tilted against each other although they are all parallel!

You can easily convince yourself of this by keeping the book just below eye level again and looking along the lines. The explanation for the tilting of the vertical or horizontal lines lies in the small alternating vertical and horizontal crossbars.

Again, our perceptual system strives to see right angles. Therefore, the continuous lines are tilted "in our imagination" away from the perpendicular of the cross struts, which causes the deception. In Fig. 2.15 on the right, you can see a very similar deception. Here the crossbeams are no longer parallel but have a gradually changed cutting angle. This deception is thus a mixture of Zöllner's and Hering's deceptions. The parallel lines now appear not only to be tilted against each other, but also bent up!

The strive for conciseness already mentioned several times, however, is only one possible explanation for the occurrence of Zöllner's deception. Another explanation

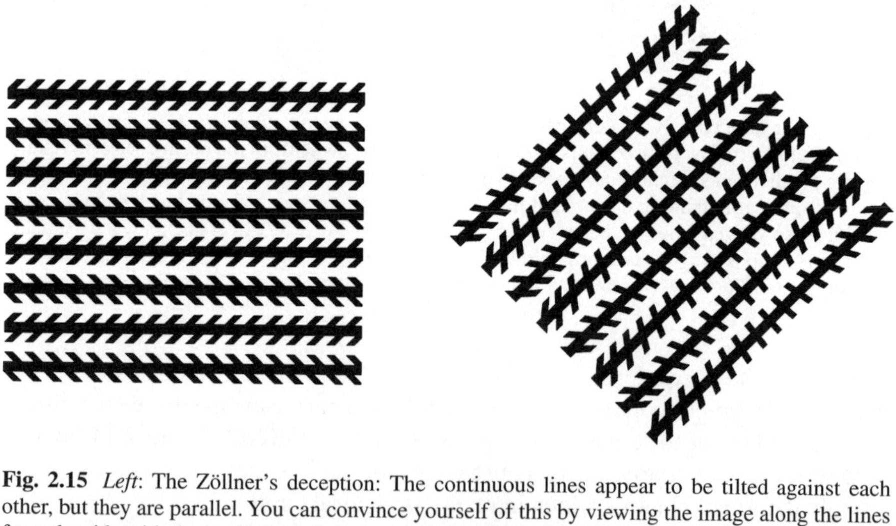

Fig. 2.15 *Left*: The Zöllner's deception: The continuous lines appear to be tilted against each other, but they are parallel. You can convince yourself of this by viewing the image along the lines from the side with the book just below eye level. *Right*: Modified Zöllner's deception: The parallel lines now appear tilted against each other and bent at the same time!

lies on the neuronal level of the detector cells. Here it is assumed that the detector cells for the orientation of lines inhibit each other if these lines have a similar orientation. This can explain the preferred perception of the tilted orientation.

A similar mechanism can easily be seen in the example of Gibson's deception, also known as the *tilting* aftereffect.

2.9 The Tilting Aftereffect

View the tilted bar grid in Fig. 2.16 left for about 30 s. Fix your view on the small grey circle in the middle of the pattern. During the 30 s you adapt the line direction. This means that the detector cells that are responsible for this orientation gradually tire!

Now look at the middle, vertical bar grid. Something very remarkable happens: You don't see these bars oriented vertically anymore—instead, they are tilted in the direction opposite to that of the adaptation pattern!

This amazing experience can of course also be made in the opposite direction. Therefore, first adapt your perceptual system to the right pattern for 30 s before you look at the vertical pattern. Again, the same phenomenon appears: The apparent angles are tilted away from the adaptation pattern (= *Gibson phenomenon*).

The detector cells for the adaptation direction are so overloaded that they no longer work as usual. This gives the perception of the other orientations an overweight, which leads to an aftereffect. This means that the image appears tilted in the opposite direction.

Gibson's deception shows that our perception tends to overestimate the angle of two bar patterns observed at a temporal distance. This effect of the inhibition/saturation of the detector cells also works naturally with simultaneous views of two clashing, differently oriented lines.

Fig. 2.16 Gibson's deception (tilting aftereffect): First fix your eyes on the circle in the left bar pattern for about 30 s. When you then look at the vertical pattern, the bars appear surprisingly tilted in the direction opposite to that of the left pattern! An exactly opposite effect occurs when you fix your eyes the right pattern for 30 s before you look at the vertical bar grid again

Fig. 2.17 Tilting circles. The circles, which are horizontally striped, appear opposite to the orientation of their environmental pattern

This is shown by looking at Fig. 2.17: The two inner circles have a horizontally striped pattern. Nevertheless, they appear tilted—exactly opposite to their environment pattern!

This tilting effect, by the way, also explains Zöllner's deception.

2.10 The Fraser Deception

On the basis of this tilting effect some very spectacular illusions can be achieved (Fraser 1908). Two of these can seen in Figs. 2.18 and 2.19.

In Fig. 2.18 it seems as if the letters are at an angle, but they are perfectly vertical. Our perceptual apparatus here is once again at its "end." Caused by the diagonal hatching, the flip-flop impression results in a way similar to that of Zöllner's deception.

Figure 2.19 shows the *Fraser spiral*. Remarkably, it is not really a spiral, but concentric rings! The hatching of the circles creates the impression that they wind inward. Try to follow the circles with your index finger or a pencil for clarification. But be careful: the illusion is so intense that it can happen easily that even our finger slips inward!

2.11 The Vertical Deception

Let's go back to Zöllner's deception:

Rotate Fig. 2.15 so that the straight lines are vertical. You will surely notice immediately that the deception has decreased considerably! This is an indication that the vertical orientation is clearly dominant for our perceptual system and reduces the extent of deception to a minimum.

A preference of our perception for the vertical orientation can be demonstrated in various ways. One possibility you will encounter in the journey to ambiguous pictures is the form of the Maltese cross; here a clear preference for the vertical-horizontal orientation can be observed.

Fig. 2.18 Fraser's deception: The letters are in reality perfectly straight!

Fig. 2.19 Fraser's spiral: In reality, the perceived spiral consists of separate circles!

Another way to demonstrate the preference of the vertical (and horizontal) directions in our perception is to measure the contrast sensitivity of line gratings. High-contrast line grids consist of alternating black and white bars. If the contrast is weakened, the black and white bars become more and more alike—and at some point a gradation is no longer recognizable.

The minimum contrast from which a bar differentiation is just still possible can be exactly measured. Thereby something important turns out: the contrast sensitivity is clearly highest with vertical and horizontal bars. On the other hand, a bar angle of 45° has the lowest contrast resolution.

This makes it clear that the vertical and horizontal orientations are preferred in our perception. Moreover, there are also considerable differences in the perception of these two directions. For example, we tend to considerably underestimate the height of a cupboard, a Christmas tree, or a construction crane as long as these objects lie flat on the ground. But if they are positioned upright, we are usually

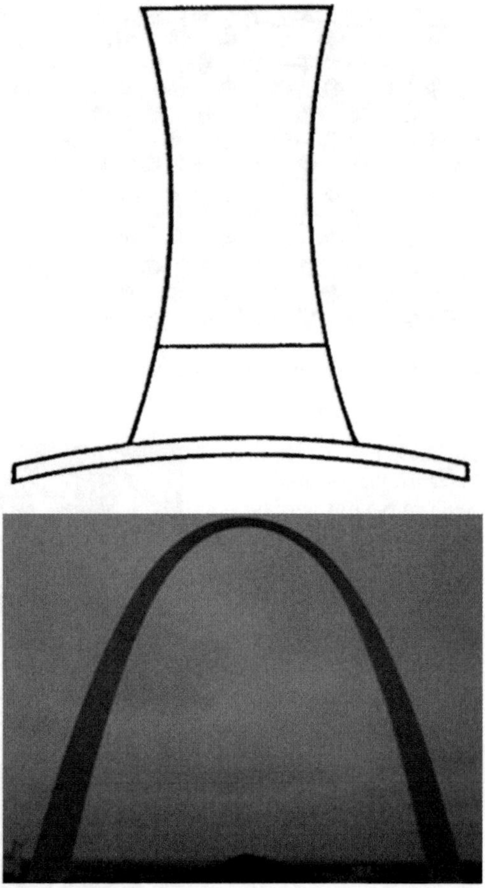

Fig. 2.20 The vertical illusion: In both images the vertical direction is overestimated by our perception. *Top*: The cylinder is as wide as it is high! *Bottoms*: The arc (Gateway Arch, St. Louis) is as wide as high! (192 m each)

surprised by their great height. The same can be observed of a factory chimney that was destroyed by a blast: The footprint of the chimney requires considerably less space on the ground than one would have expected if one had not measured.

Would you, for example, have noticed at a superficial glance that the cylinder in Fig. 2.20 is as wide as it is high or that the Gateway Arch in St. Louis, displayed at the bottom, is exactly as wide as it is high?

This overestimation of height in comparison to width is probably due to the nature of man, who has been firmly bound to the ground by gravity for millions of years. To cover a kilometer of altitude is incomparably more difficult for a human being than to cover a kilometer at ground level. Therefore, it is only understandable that the vertical is also constantly overrated in our perception.

Fig. 2.21 Length illusion, evidenced by different hatchings: The square on the left appears wider than normal due to the vertical stripes. Conversely, the horizontal stripes make the same square appear narrower on the right! With these stripes on a T-shirt you can correct every belly circumference—if not really, then at least optically!

2.12 T-Shirts, Cross- and Longitudinal-Striped

The misinterpretation of height and width can be intensified by clever hatching—see Fig. 2.21 for more details.

The dimensions of the two-line arrangements are exactly the same. Nevertheless, the left, vertically striped pattern appears astonishingly wider than the right, horizontally striped pattern, contrary to widespread popular opinion! Actually, the vertical illusion underestimates the horizontal lines compared to the vertical lines. This deception, however, is overruled here by another, very intense effect: The length impression of a line is clearly enhanced by crosswise bars (Oppel-Kundt's deception, see also Fig. 2.22).

Fashion designers and manufacturers make clever use of this knowledge for their customers. Thicker people should wear T-shirts with horizontal stripes to look slimmer. Conversely, vertical stripes are worn to appear wider! At the moment, cross-stripes are very "in"—so one more hamburger can be eaten without problems.

2.13 The Oppel-Kundt Deception

Which distance in Fig. 2.22 do you perceive to be shorter? You probably chose the disctance without vertical intermediate lines, although you already suspected that the two disctances are exactly the same length again! The vertical intermediate bars widen the length impression of a track—it's *Oppel-Kundt's deception!*

An underestimation of "empty" rooms occurs quite frequently. We only really become aware of the size of an empty room when it has been furnished. The underestimation of empty rooms also ensures that, contrary to an initial estimate of proportion, it is usually possible to bring a sofa through a door or place a cupboard in a supposedly too-small place on the furniture van.

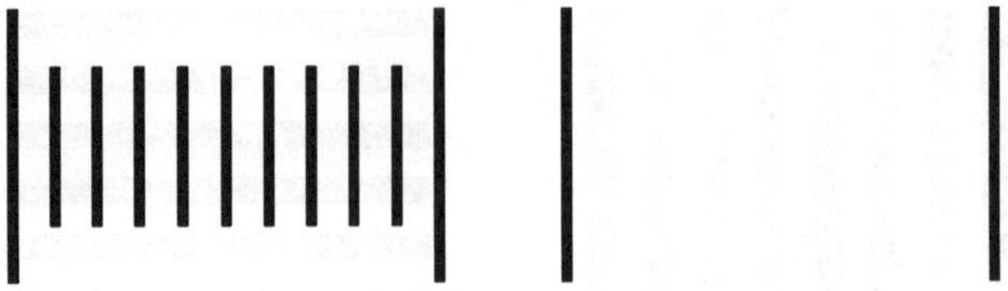

Fig. 2.22 Oppel-Kundt's deception (1895): The left distance with the vertical auxiliary lines appears much longer than the left, but equally long distance on the right

An explanation of this deception can be seen in the unconscious spatial interpretation of the image. The vertical intermediate lines let us look further into the depth. The misinterpretation of the size constancy results in the overestimation of the left distance, while the "empty" right distance is correspondingly underestimated.

2.14 The Moon Deception

Another consequence of this underestimation of the "empty" space is the famous *moon deception* (Fig. 2.23). This deception has been known for centuries and there is a multitude of theories and books to explain it.

Perhaps you have already noticed that the moon looks much bigger when it stands on the horizon than when it stands directly above you in the sky. Measurements by Kaufman and Rock (Kaufman and Rock (1962a, b), Rock and Kaufman (1962)) showed that this deception causes a size difference of more than 30%. From these investigations it also resulted that the different size impression of the moon is determined by the comparison with the horizon.

If, for example, the horizon is faded out by means of an aperture, the moon at the bottom appears in the same size as the moon at the zenith. The moon at its zenith does not have any comparison points for distance estimation. However, as we already know, the "empty" space around it causes its distance to be greatly underestimated. On the other hand, a distant horizon makes the moon appear farther away.

With this, the misinterpretation of the size constancy is again outdistancing us. The moon disc on the horizon, which appears to be farther away, is automatically perceived as larger than the moon disc on the sky scene, which appears to be closer.

Our perceptual system has been designed to detect and amplify differences in the structures of our environment. This property enables us to recognize patterns and to distinguish them from one another. For example, our perception has developed mechanisms through which it is able to distinguish between minimally different bright or colored surfaces. We will get to know these "recipes" of brightness contrast enhancement and color contrast enhancement in detail during the journey through form and figure as well as the journey through color vision.

Fig. 2.23 Representation of the Moon—deception: on the horizon the moon appears about 1.3 times as large as on the sky scene

2.15 The Titchener Deception

Our perception can increase not only brightness contrasts, but also size contrasts. An example of this can be seen in Fig. 2.24—the *Titchener deception.* A test object is perceived differently in size depending on the size of its environment: This leads to a significant increase in size contrast. If the environmental patterns of the test figure are large, the test object is perceived as small. Conversely, the test figure appears larger in a smaller environment.

Compare the size of the two middle circles in Fig. 2.24. The upper circle is in the vicinity of large circles; it therefore appears much smaller than the identical lower circle, which is surrounded by many small circles!

After these numerous deceptions, you have earned yourself a small snack. That's why we quickly find our way to the nearest café, which bears the strange name "Double Vision Bar." Hungry, you order a cake as big as possible and a banana.

2.16 The Ebbinghaus Deception

In the meantime, the waiter, who also looks a bit odd, gives you two cake tins to choose from. On both cake trays there are still three pieces of cake, as you can see in Fig. 2.25.

"Pick the biggest piece of cake, but only choose between the two middle pieces," he says.

You will probably choose the left middle piece. The waiter is pleased, because you have decided in favor of the smaller piece! The two pieces of cake are a further example of the strong impact of the size contrast enhancement effect: this is the *Ebbinghaus's illusion.* The middle piece of cake on the left-hand side seems to be considerably larger than the piece in the middle on the right-hand side, which is actually slightly larger!

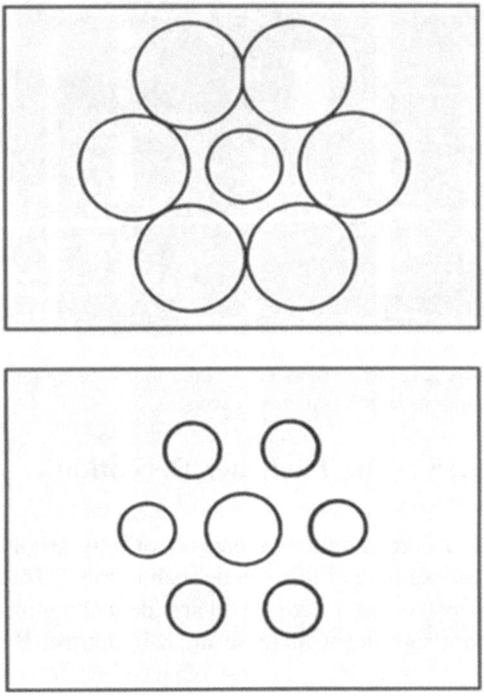

Fig. 2.24 Titchener's deception (1898): Compare the two equally sized middle circles with each other! What do you notice?

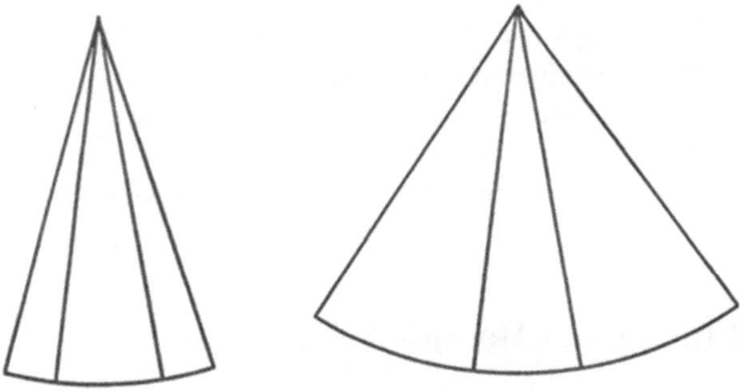

Fig. 2.25 The two cake trays—this picture is an example of *Ebbinghaus size contrast*: The piece of cake in the middle of the left tray appears considerably larger between the two narrow cake pieces than the right cake middle piece between two large pieces, although in reality this one is slightly smaller!

2.17 The Jastrow Size Illusion

Meanwhile the waiter has returned with two bananas. "Again, I want you to choose the bigger one!" The bananas are cut off vertically on both sides and lie on a silver tray, similar to Fig. 2.26.

Fig. 2.26 The two bananas: The upper banana is actually bigger than the lower one! This impressive size deception is called Jastrow's deception after its "inventor"

Fig. 2.27 The supposedly inclined trays are really parallel!

You will probably not hesitate long and decide for the much bigger banana below. Again, the waiter is pleased, because in reality the upper banana is the somewhat larger one! This impressive banana delusion is the *Jastrow size illusion*.

2.18 The Trick with the Trays

The waiter is now busy removing two stacks of trays. These seem to lie extremely crooked, which you can see for yourself in Fig. 2.27. Strangely enough, he doesn't seem to have any problems with carrying them away. That's no wonder when you take a closer look, because the trays are actually aligned exactly parallel!

In amazement we decide to sit down for safety reasons. On the chair there is a high and comfortable cushion (see Fig. 2.28), and we sit down quickly on it. But once again we are mistaken—and bang hard on the flat chair seat.

After the laughter of the waiter has subsided, he says: "Similar to the deception in Fig. 2.27, the lines in Fig. 2.28 are not curved but parallel. The basic pattern is a completely symmetrical checkerboard pattern with right angles. Only through the additional small squares a spatial geometry is impressed on the whole pattern, which makes the picture appear as a puffed-up cushion. You can convince yourself of the parallelism of the basic pattern lines by looking at the sheet glancing off from the side."

The tour group looks at the waiter with big eyes—as big eyes as seen in Fig. 2.29. You won't be surprised about anything anymore: In this picture the lines are all

Fig. 2.28 The cushion of Kitaoka. (With kind permission of Akiyoshi Kitaoka)

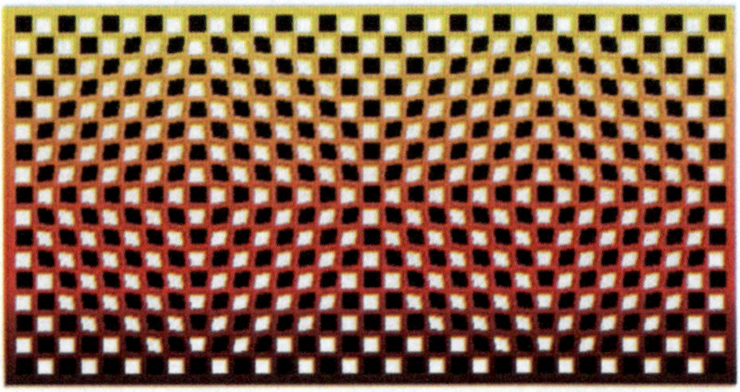

Fig. 2.29 Eyes. (With kind permission of Akiyoshi Kitaoka)

parallel. The spatial curvature is merely illusory, created by the orientation and geometry of the black and white surfaces. If you look at the picture flat streaking from the side, the illusion dissolves.

To be on the safe side, we order the bill to get out of the bar as soon as possible. The strange waiter seems to have been waiting for it and brings us a bill immediately. As can be seen in Fig. 2.30, it consists of a few narrow beams—nothing more can be seen on them!

Everyone is puzzled, but there is no solution in sight—until the bill is blown to the ground by a gust of wind. When we pick it up, our table neighbor looks at the sheet diagonally from below and suddenly shouts: "I got it!" In fact, you can see the numbers on the invoice in Fig. 2.30 very well if you look flat over the sheet from the front again. It's the perspective that counts!

Fig. 2.30 The bill

Quickly we put 17, 50 $—without tips, of course—on the table and leave the café, before the strange waiter comes with two different sized purses or comes up with something new to fool us.

2.19 The Sunset

On the journey back home in our coach through the wonderland of perception we enjoy a wonderful sunset. Suddenly a narrow band of clouds moves in front of the sun (Fig. 2.31).

You probably don't wonder about anything anymore—especially not about the optical illusion that occurs as a result: the sun appears to be strongly squeezed. It looks a lot taller than it is wide!

Fig. 2.31 A view from the bus window: The sun behind a bank of clouds appears higher than wide!

Soon thereafter, however, this riddle solves itself. Because—you have already guessed—the sun is neither high nor wide.

The sun ends its daily work and our exciting journey to the fantastic land of optical illusions.

Chapter 3 Third Journey: Perception of Forms and Brightness

Trails that cross each other are an occasion to pause—they raise the question of where from and where to, and of the deeper meaning of space, time, and goal. How sure can we be that the current path we have chosen is also the right one for us?

Completely regardless of the sheer enormous variety of possibilities, each path leads to a specific goal. The signs that line the wayside want to be interpreted critically and with an alert sense—this is the only way to open the view for clarity and reality.

This third journey of seeing and perception is as much about such signs to be interpreted as it is about (more or less) clear forms, perfectly formed figures, and fascinating illusions with wonderful contrasts.

3.1 Reflected Light

In this journey we follow the traces of two of the most basic principles of vision: the perception of brightness and the perception of form. The starting point of the journey is a motorway bridge in bright sunshine. As we already know from our first journey, the sunlight is reflected to varying degrees by the objects around us, for example the hats of our fellow travelers, their clothes, and their cameras. A part of this reflected light finally reaches our eyes.

A dark camera, for example, appears dark because very little light enters our eyes from it; rather, it absorbs almost all the incident light. In contrast, a white hat absorbs hardly any light; it almost completely reflects the light rays. In other words, the hat transmits intense rays of light to our retina, causing us to perceive the object as bright. We can hold firm: The more the light is reflected, the brighter the surface appears to us.

Surely you have worn a black T-shirt in summer. Then you have probably also noticed what happens to absorbed sunlight in dark fabrics: Its energy is converted

T. Ditzinger, *Illusions of Seeing*, https://doi.org/10.1007/978-3-030-63635-7_3

into heat. The logical consequence is then: In a dark T-shirt you sweat more than in a light one.

This effect can also be felt on the bridge over the motorway car park, where there is only some shade at the edge. However, the bus driver has already occupied this space with a large number of paint pots. While he begins to paint parts of the asphalt with a large brush, he tells of a strange dream he had last night.

3.2 The Dream of the Bus Driver

In this dream the sheep from the first journey appeared to him again; they again formed various new formations. The first of these can be seen from a bird's-eye view in Fig. 3.1.

3.2.1 Symmetry

If you look at the figure, you probably have the impression that the symmetrical black areas are in the foreground—regardless of their shape.

This already makes it clear that symmetry is a very powerful law of seeing. This becomes particularly obvious through several investigations by Bela Julesz, which are illustrated below (Julesz 1971, 1986).

It is well known that our perception strives for order and concise forms. This principle of conciseness is fulfilled to a high degree by a recognized symmetry. So we can recognize symmetrical structures even in pictures with seemingly completely random patterns. You can easily check this by looking at Fig. 3.2: Two symmetry axes are clearly visible, which meet in the middle of the picture!

Fig. 3.1 The dream of the bus driver, first picture: the influence of symmetry on figure perception. The symmetrical surfaces seem to be in the foreground—regardless of their brightness!

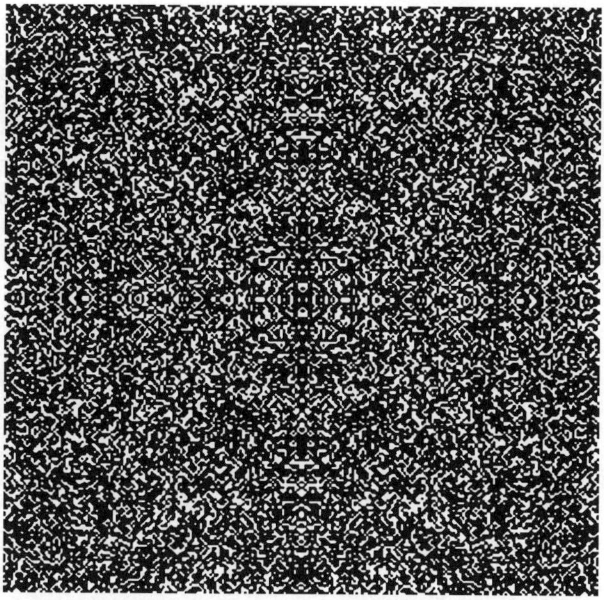

Fig. 3.2 Do you recognize the symmetry? The two symmetry axes meet in the middle of the picture!

The image was created as follows: First, the upper left quarter of the image was randomly patterned black-and-white. Then this image quadrant was mirrored downwards. In the last step, the left half was mirrored to the right.

Our perceptual apparatus has to continuously process an enormous amount of information. To avoid being overloaded, it tries to work as economically and with as little effort as possible. This successful recipe of the simplest procedures can be experienced in the next two figures. They show that the recognition of symmetries works by means of an amazingly simple trick.

Figure 3.3 shows the same symmetrical image as Fig. 3.2, but with one small difference: Along the symmetry axes, the original pattern was replaced by a different random pattern in a width of eight image pixels (the smallest picture point is an image pixel). This minimal difference has drastic effects on our perception: We are no longer able to recognize the symmetries of Fig. 3.3 as such! This at first very surprising result suggests the assumption that our perception only compares the immediately adjacent pairs of points when recognizing symmetries.

This assumption can be verified very easily by the reverse test:

Consider Fig. 3.4 for the crosscheck. You will certainly recognize the two symmetry axes that are already familiar to us from Fig. 3.2, intersecting in the middle. In reality, Fig. 3.4 is not symmetrical at all! It's an entirely random pattern. The symmetrical pattern from Fig. 3.2 was inserted only in a width of eight pixels around the symmetry axes. These few symmetrically arranged pixels are enough to make the whole picture appear symmetrical!

With this "economic trick," our perceptual apparatus accepts the possibility of a misinterpretation of the entire image—but it avoids the danger of overload. In order to correctly examine the image for symmetry, our perception would have to compare

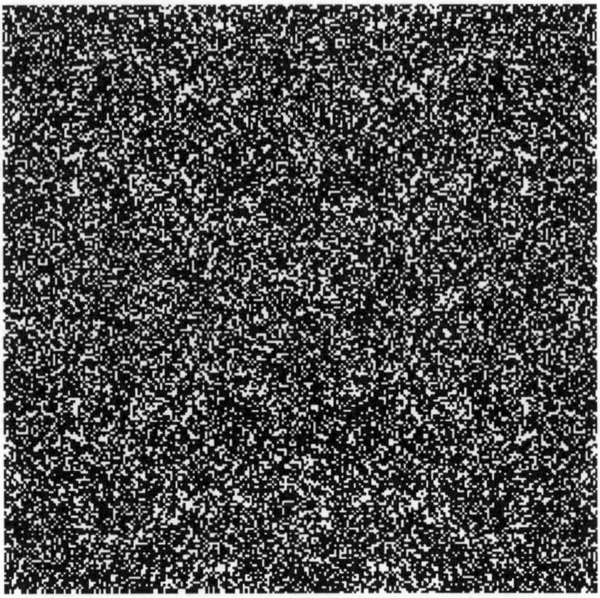

Fig. 3.3 A symmetrical image that does not appear symmetrical! Only on the symmetry axes intersecting in the middle another random pattern was built in in a width of eight pixels

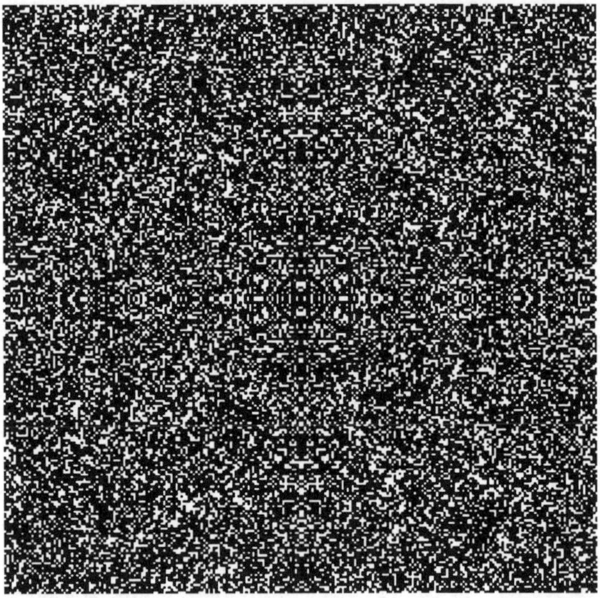

Fig. 3.4 An asymmetrical image that appears symmetrical! This picture is the exact counterpart of Fig. 3.3. It consists of a completely random pattern. A symmetrical pattern was built in only on the axes that intersect in the middle

each individual pixel with each other. With the number of 57,600 pixels as in our picture with 240 points length and 240 points width, that would be an astonishing 1.659 billion comparisons—and all of that only to check the symmetry properties of a single picture! By using the trick of checking only a few neighboring pixels for their symmetry instead, our perception has found a very pragmatic compromise.

Fig. 3.5 The dream of the bus driver, second picture: The law of proximity overrides symmetry. The symmetrical patterns now seem to lie in the background

Nevertheless, symmetry is not the only decisive factor in finding a concise form. So further grouping laws, which you already got to know in the first journey, can step into a competition with the law of symmetry. This can be confirmed by the next sample formation of sheep from the dream of the bus driver (Fig. 3.5).

3.2.2 Proximity Against Symmetry

How this "competition" of the two opposing Gestalt laws ends can be seen very clearly from looking at the patterns in Fig. 3.5. The surface with the good shape seems to be in the foreground; this foreground surface is called the *figure*. The less concise shape disappears into the background. In Fig. 3.5, the law of proximity overrides the property of symmetry. The symmetrical surfaces no longer appear in the foreground here.

The distance between the surfaces can be easily changed, which means that the influence of the law of proximity has less effect. The "competition" of the Gestalt laws can therefore be arranged to result in a draw—as shown in Fig. 3.6.

Our perception gets into a great conflict through such arrangements. Our perception knows no such draw and is designed to always have only one content of consciousness! Thanks to its ingenuity, however, perception succeeds in finding a way out of this situation: It simply alternates between the two equally good figures!

You will get to know this fascinating constellation even better in the fourth journey into the world of ambiguous perceptions.

3.2.3 Closedness Against Symmetry

Now look at the last dream image of the bus driver (Fig. 3.7). In this picture you can see another example of two competing Gestalt laws: Closure against symmetry.

The *law of closure* is also called *law of the inside* or *law of convexity*. The special feature of the black surfaces in Fig. 3.7 is that they are mainly curved and rounded outwards.

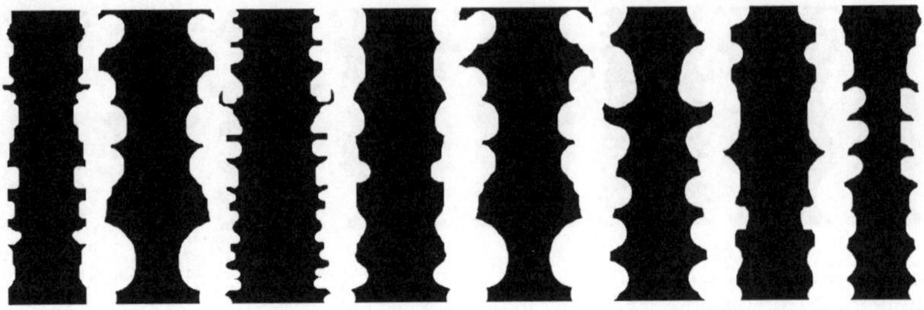

Fig. 3.6 The dream of the bus driver, third picture: The laws of shape of proximity and symmetry are approximately equally strong. Our perception cannot come to terms with a draw and tries to alternate between the two figures!

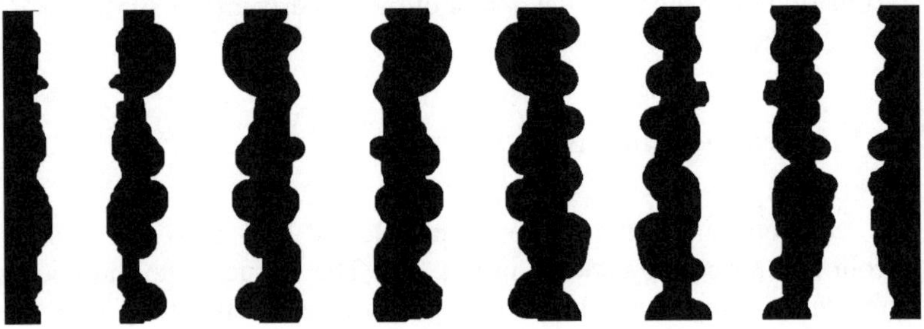

Fig. 3.7 The dream of the bus driver, fourth picture: The convexity overrides the symmetry. The convex black surfaces clearly appear as a figure, while the symmetrical white surfaces seem to lie in the background

In Fig. 3.7, this convexity of the black surfaces clearly has a perceptual predominance over symmetry. Thus, the asymmetrical but convex black surfaces seem to be in the foreground and the symmetrical white surfaces seem to be in the background.

3.3 The Motorway Bridge and the Asphalt Figures

You have seen through many of the images we have looked at so far how difficult it must be for our perceptual apparatus to weigh all the laws of grouping against each other and finally interpret the most concise form as a figure. With some pictures it seems at first even quite impossible to recognize a figure. This was the case, for example, with the cow in Fig. 1.5 in Dallenbach's figure. Only the recognition of a meaning in terms of content enables us to see parts of the picture as a figure.

3.3.1 A Block Diagram

The bus driver pulls a heavily crumpled picture out of his trouser pocket, which he brought with him especially for the guests of his adventure trip. It is a block diagram produced with a simple raster technique.

With this method, an image is first selected in normal resolution, over which a very coarse raster is laid. This splits the image into individual blocks. A uniform hue is assigned to these blocks, namely the respective mean value from those picture elements that lie within the block. This results in a pattern as shown in Fig. 3.8, in which you will probably initially recognize only randomly distributed rectangular areas.

The difference in brightness, i.e., the *contrast* between adjacent blocks, is sometimes very high. In contrast, the normal resolution image has gradual transitions and lower contrast. This difference must be compensated for when recognizing block diagrams.

It is best to pinch your eyes together a little, darken the light in the room, or view the picture at a greater distance. All these methods help our eyes to blur the high-contrast block boundaries somewhat.

This visual technique is the exact opposite of what we normally do when we look at a picture: As a rule, we focus our eyes as much as we can to see as many image details as possible. This new visual technique makes Fig. 3.8 appear much more natural. It's not too difficult now to identify the familiar face. At this moment of recognition something very decisive happens in our shape and figure recognition: Suddenly certain blocks form a figure, while others move into the background. The grouping law of content meaning has here taken over the organization of figure perception.

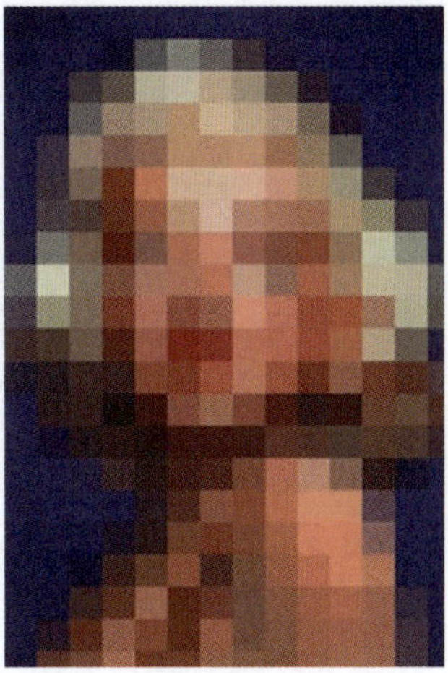

Fig. 3.8 A block diagram: The seemingly random arrangement of blocks of different brightness suddenly results in a familiar face! There is a paradoxical trick with which you can help your eyes on the jumps. To see the picture, you must squint your eyes, observe the picture from a greater distance, or darken the light in the room. The resolution is shown in Fig. 3.14

3.3.2 Simple Brightness Illusions

In the meantime, the bus driver has painted works of art in various colors on the asphalt. He explains the following to the passengers: "These asphalt drawings are intended to show you how our perception of brightness works. The perception of brightness is by far not as simple and clear as some of you might think. Brightness is much more than just measuring the strength of reflected light rays.

The impression of brightness of a surface depends strongly on its surroundings. In front of a dark background, a bright surface looks much brighter than in front of a bright background. Please take a look at the white parking lot markings in Fig. 3.9!"

The bus driver continues, "I painted the black asphalt around the markings from left to right with increasingly lighter paint. Now you are sure to notice that the marking area, which in reality is always the same brightness, is becoming increasingly grey!"

After a short break, the bus driver continues his explanation: "This brightness illusion becomes even more impressive with the second asphalt work of art (Fig. 3.10). I have painted the background of a constant grey marking area from left to right continuously darker—which, by the way, was a very exhausting painting job. Concentrate on the horizontal strip, please. Surely it seems to become brighter and brighter from left to right! In reality, however, it is uniformly grey! You can easily convince yourself of that by covering the background with your hands."

"To sum up," says the bus driver, "the brightness sensation is always relative. Depending on the brightness level of the environment, a physically unambiguous brightness tone is reinterpreted. The brightness differences of adjacent surfaces are exaggerated. This perceived amplification of differences in brightness has the same effect on both surfaces involved. The lighter of the two areas is further brightened, while the darker one appears even darker."

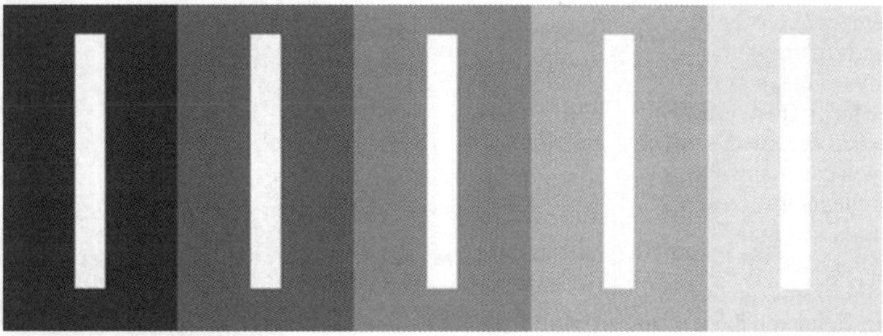

Fig. 3.9 The view from the motorway bridge: The white parking lot markings appear increasingly darker from left to right, although in reality they have the same brightness level

Fig. 3.10 The horizontal stripe is actually uniformly grey!

Fig. 3.11 Influence of the surface shape on the brightness illusion. The background appears increasingly brighter from left to right – although in reality it is of the same brightness. However, this brightness deception is much weaker than the opposite case in Fig. 3.10

3.3.3 More Complicated Brightness Illusions

The contrast we perceive is very much dependent on the type of surface, the nature of its boundary lines, and the distance between the visual elements.

The brightness deception in figures consisting of small areas, as in Figs. 3.9 and 3.10, is very pronounced. On the other hand, the brightness deception is less clear with surfaces that appear in the background. You can check this in Fig. 3.11: The shapes are exactly the same as in Fig. 3.10. However, the background now has a uniform gray tone throughout. The crossbar, on the other hand, becomes increasingly darker from left to right. We notice: The influence of this color gradient on the background is present (at the right edge of the image the background appears a little brighter), but much less clear than in the opposite case of Fig. 3.10.

Fig. 3.12 Koffka rings: The ring has a uniform brightness in tone, but appears brighter on its left half and darker on the right. The more the two halves of the ring are separated from each other, the stronger this contrast illusion becomes

To explain further factors influencing the perception of brightness to his guests, the bus driver has now drawn five different versions of the so-called *Koffka ring* onto the road. The ring, named after the Berlin Gestalt psychologist Koffka (1886–1941), consists of two halves that are uniformly bright. The two halves of the ring are each underlaid with a dark background (left) and a light background (right).

In Fig. 3.12, upside left, the ring is separated from the background by a black line. I'm sure you will notice the contrast illusion: On the left, the ring appears much brighter than on the right. This contrast illusion disappears almost completely if the separating line is omitted—you can check this in Fig. 3.12, top right: On the left the ring appears subtly brighter than on the right!

This effect can be explained by the fact that our perception apparatus largely softens the strong contrast illusion caused by the background.

Figure 3.12, bottom left, shows the influence of the different nature of the boundary lines. The two halves of the ring are separated from each other by serrated lines and set off from the background. The jagged edge reduces the contrast effect of the background and the illusion does not appear as intense as in the picture above left. This shows that the clearer the area boundary, the stronger the contrast illusion. If the two halves of the ring are completely separated from each other as shown in Fig. 3.12 on the lower right, the illusion is even stronger than on the upper left.

The size of the figures is also an important factor in causing the overestimation of the brightness differences. The contrast illusion is slightly stronger with a small Koffka ring, as in Fig. 3.13, compared to Fig. 3.12 on the upper left.

Fig. 3.13 The smaller the ring, the stronger the contrast illusion!

Fig. 3.14 This is behind Fig. 3.8!

Why does brightness contrast enhancement exist and how can it be explained biologically?

The main task of our perceptual apparatus is to recognize structures that consist of flat patterns. However, since the brightness of neighboring areas is often quite similar, it is not easy to spot a lizard on a stone, a hare in the forest, or a polar bear in the snow.

Our senses are therefore designed to amplify differences in brightness between adjacent surfaces. That is: A bright surface is lightened by our perceptual apparatus in the direction of the border line, while the darker one is darkened.

Fig. 3.15 The Mach Bands: The Mach bands (Mach 1914) appear at the boundaries of the surfaces with uniform color. On their left half they appear brighter than the uniformly colored surface, while on the right they appear darker!

You can see this for yourself on the next asphalt drawing (Fig. 3.15). This probably best known contrast illusion, is attributed to the Austrian physiologist Ernst Mach.

3.3.4 The Mach Bands

The Mach illusion is a sequence of stripes that become increasingly brighter from left to right. Within the stripes, however, the degree of brightness is exactly uniform. In the area of the transition edges, the impressive vertical *Mach stripes appear*. These make the left side of a uniformly colored stripe appear brighter and the right side darker to the observer.

As we know today and as you were able to determine yourself during your first journey with the help of Fig. 1.3, the photoreceptors of our retina gradually pass on the sensory impressions to subsequent cell layers, which summarize and process this information. The first layer are the bipolar cells, the second the ganglion cells. However, a photoreceptor does not only forward its information to a single downstream cell "upwards." Rather, such stimulus transmission also occurs "horizontally," i.e., many different processing cells are interconnected. The processing of a visual stimulus takes place in very different ways, depending on the nature of the stimulus.

As early as 1860, Mach suspected that neighboring photoreceptors were spatially connected in a very sophisticated way. All photoreceptors in the center of a particular area, the *receptive field*, are "equal."

The brightness stimuli they absorb are added together in the cell responsible for this receptive field.

The photoreceptors, however, which are located in the outer areas of the receptive field, act exactly the opposite way on the responsible calculation cell. This means that the stimuli absorbed by them are subtracted from the calculated brightness level. This spatial summary is also called *the principle of lateral inhibition*. With this wonderfully simple trick, the fantastic visual performance of the brightness contrast enhancement can be explained.

The ability to recognize different degrees of brightness is a prime example of a recipe for success in our perception. We do not always perceive the true physical brightness conditions of our environment; on the contrary, the perception of brightness sometimes does not have too much to do with reality, as the asphalt drawings we have already seen prove. Through the mental reinterpretation of the actual brightness values, we are able to clearly distinguish surfaces from each other and see them sharper.

But how does this complicated visual performance come about? The answer to this question leads us to the roots of the human perception of brightness. Our nature only needs a single, ingeniously simple trick for this: lateral inhibition (which was already mentioned) is an important property of every single photoreceptor. The interaction of several photoreceptors finally enables the overall superior visual performance of edge contrast enhancement.

The photoreceptors located on the retina transmit visual stimuli to the subsequent cell layers, where neurons evaluate the signals differently according to their origin. The incoming visual information is networked in such a way that it is spatially combined into a receptive field.

Each photoreceptor cell is associated with many receptive fields. Conversely, every receptive field is associated with many photoreceptor cells. The stimuli of photoreceptor cells from the central area of the receptive field are evaluated differently than those of photoreceptor cells in the outer areas. There are different evaluation neurons with different evaluation methods. In the *on-center neurons*, stimuli from central areas have an activating effect, while stimuli from outer areas have an inhibiting effect. Exactly the opposite is true for *off-center neurons*.

How does this elementary spatial connection result in such a general system property as contrast enhancement? This becomes clear when you look at Fig. 3.16—here the calculation process in a receptive field is reproduced when looking at a brightness transition in an on-center neuron—although somewhat simplified: On the left there is a white stripe with the imaginary brightness 10 and on the right a dark stripe with the brightness 0.

The receptive field is represented by a circle. The inner circle represents the center of the field.

What is the result of our calculation if the receptive field lies completely on the white stripe? This case is shown in Fig. 3.16 on the left: The photoreceptors of the center convert the received brightness into an excitation strength with the intensity 10. The photoreceptors in the outer area also receive the white color, but have an inhibitory effect on the receptive field. We therefore assume an effect of minus 1 from four directions on the overall result. This is schematically represented by the four numerical values in the outer circle. The final excitation of the on-center neuron results from the addition of the values of the center and the outer area, which leads to an activation of intensity 6.

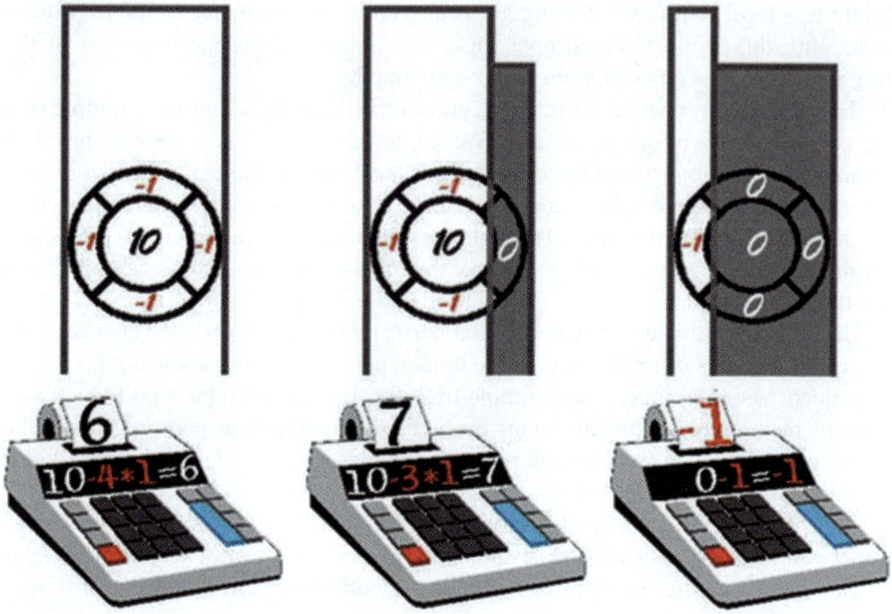

Fig. 3.16 Overview of the "data center" of a brightness evaluation neuron

In the middle of Fig. 3.16 a receptive field is shown that registers a black area in the right outer area. This leads to less inhibition of the receptive field in this range (0 instead of 1). Since all other conditions have remained unchanged, a higher activation of the neuron results! And it is precisely this fact that causes the edge contrast enhancement. Although darker light is incident, brighter light is registered.

Figure 3.16 on the right shows the case where the center also registers black light. Therefore the areas with the intensity 0 are excited—only not the left outer area. The overall result is even blacker than black due to the lateral inhibition from the outside left. Only if the whole receptive field sees black, the total excitation is 0.

For the off-center neurons, which are connected in exactly the opposite way, the function of the lateral inhibition can be demonstrated in the same way as in Fig. 3.16—albeit with a negative sign.

All contrast enhancement effects can be explained from this simple example. White becomes brighter and black is perceived darker when one looks at their common boundary.

3.3.5 The Craik-Cornsweet-O'Brien Deception

The *Craik-Cornsweet-O'Brien illusion* (O'Brien (1959), Craik (1966), Cornsweet (1970)) proves that contrast enhancement takes place in the vicinity of border lines between different surfaces. In Fig. 3.17 you surely can immediately see two

Fig. 3.17 The Craik-Cornsweet-O'Brien illusion: Certainly, you recognize two different surfaces, the left one looking darker than the right one. In reality, the two surfaces are exactly the same except at the cutting edge! Convince yourself of this by covering the area of the cutting edge with a finger or pencil

surfaces with different brightness, which can be clearly distinguished from each other; the left surface appears darker than the right one. In reality, however, it is an optical illusion—again!

To unmask the deception, simply close the transition surface with your index finger. You will immediately notice that the two surfaces are absolutely identical in their brightness level. The left surface is only darker shortly before the dividing line. On the other hand, the right surface is correspondingly brightened shortly before the dividing line.

How does this deception come about? Our eyes first notice the edges and intensify the edge contrast at the two borders accordingly. This creates the astonishing (false) impression of two different grey areas.

3.3.6 The Hermann Grid

The edge contrast enhancement is the trigger for another remarkable deception, which was discovered by chance by the physicist Hermann in 1870 and can be seen in Fig. 3.18. In *Hermann's grid* (Hermann 1870), many black squares are arranged in equal distance from one another. The spaces between the black squares are white.

But they don't appear that way everywhere!

At the intersections of the black squares, small black dots flicker! These actually white intersections appear darker because there is no direct contact to the black surfaces. A marginal contrast enhancement cannot therefore take place here to the same extent as in the other white areas. The white of the crossing points is not further brightened by the lateral inhibition, and therefore appears darker. The flickering is caused by the fact that our eyes never look exactly at the same point for very long, but are subject to a constant movement of their own. This movement represents a kind of protection against a supersaturation of our photoreceptors by receiving always the same light.

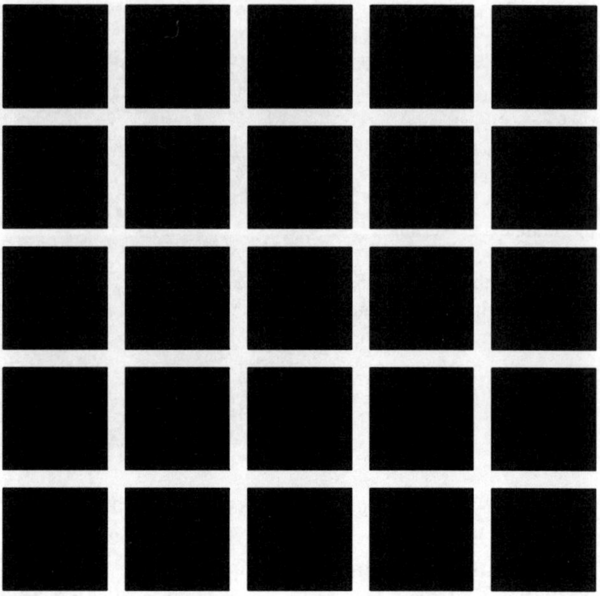

Fig. 3.18 Hermann's grid: Black flickering dots appear at the intersections of the black squares. The white of the crossing points is not further brightened by the lateral inhibition and therefore appears dark

Try to focus your view on a black flicker point in Fig. 3.18. You will notice that this then disappears immediately. The other flicker points you look at out of the corners of your eye will stay intact, though.

How can this phenomenon be explained? The answer to this question can be seen from a close look at Fig. 3.19. There, a modified Hermann lattice can be seen, with the thickness of the white lines increasing from top left to bottom right. Please focus on the white crossing points at different places. Surely you will notice a clear difference from top left to bottom right. At the top left, the black flicker points remain visible even if you fix it centrally. This will be more difficult for you farther down on the right in Fig. 3.19. To maintain the black flicker there, you need to look at the white intersection much more peripherally, out of the corner of your eye!

The explanation for this different behavior must therefore be found in the spatial conditions of the retina. In the area of the sharpest vision, the fovea, the photoreceptors lie very close together. This results in very narrow receptive fields. On the other hand, the receptive fields in the outer area of the retina are becoming larger and larger.

Hermann's deception is only possible if the expansion of the receptive field is not too small compared to the width of the white bars. The size of the receptive fields must therefore match the size of the intersection points. The illusion with central fixation and narrow receptive fields only works with very narrow bars. The more peripher the crossing points are viewed, the more easily the black dot deceptions can also be seen in the area at the bottom right of Fig. 3.19. Figure 3.19 can thus be used very excellently for the size measurement of receptive fields.

These different geometrical conditions are very nicely summarized in the circular variation of the Hermann grid by Akiyoshi Kitaoka (Fig. 3.20). In the interior of

Fig. 3.19 Experiment with Hermann's lattice: From top left to bottom right the thickness of the white bars increases. Only in the upper left corner you will succeed in maintaining the illusion if you look directly at the intersection point! Further down you have to watch the flicker points more from the corners of your eyes to see them

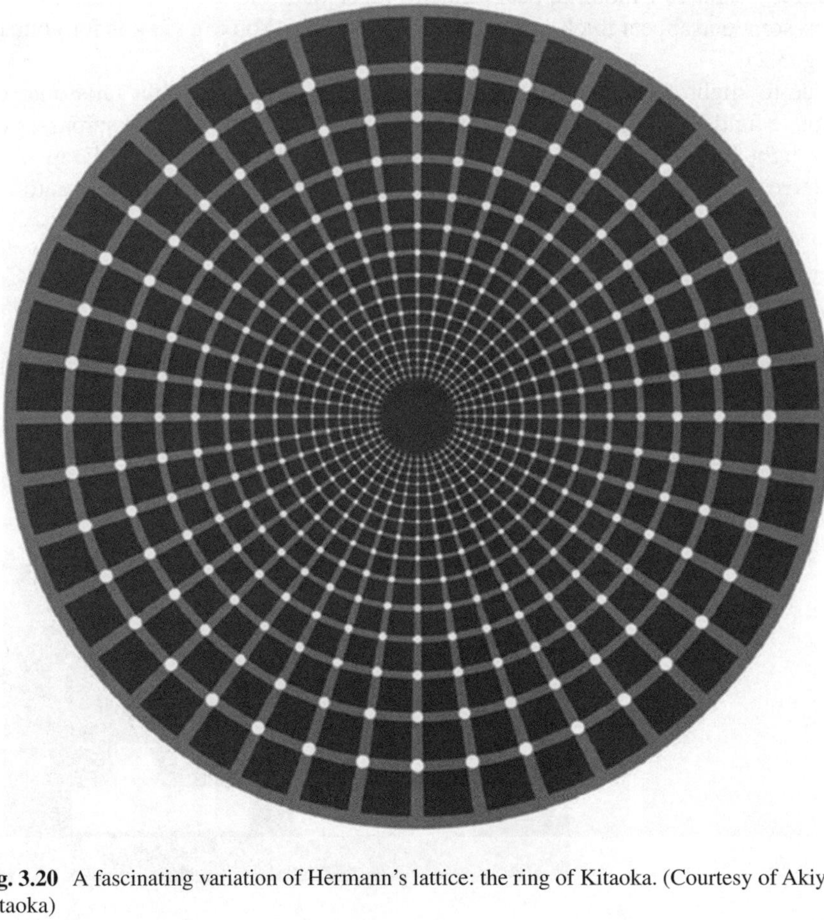

Fig. 3.20 A fascinating variation of Hermann's lattice: the ring of Kitaoka. (Courtesy of Akiyoshi Kitaoka)

the circle, the intersection points are very small and the image appears to be reasonably stable. If one looks further into the outer areas of the circle, the picture gains a lot of movement.

The intersection geometries are now wider and the dark, flickering intersection coloration occurs in peripheral vision. As soon as you look at this point centrally, it disappears again.

The situation is similar like for a shark swimming within a swarm of small fish. The small fish are represented by the moving black spots and you are the shark – or better your visual system. Wherever the shark/your view goes – the fish quickly move away!

3.3.7 The Irradiation

The slight movement of the eyes also provides a possible explanation for the effect of *irradiation.* Johann Wolfgang von Goethe and the astronomer Tycho Brahe already reported that light objects appear larger than dark objects of the same size. Thus the bright full moon appears much bigger than the dark and light clothing makes someone appear thicker than does dark clothing. You can see this for yourself in Fig. 3.21.

Due to slight eye movements, alternating white and black light falls onto the receptive fields in the edge area. The white light is much brighter and surpasses the black light by far. It leaves clear traces in our perceptual apparatus. Even if the receptive fields already have a different position, it takes some time until the

Fig. 3.21 The irradiation. Top: A white area appears larger than a black area of the same size. Bottom: The single white square in the middle seems impossible to fit into the black gaps, although it has exactly the same size

excitement caused by the white light subsides. This decay phase ultimately results in the apparent enlargement of the white areas.

3.3.8 *Bright and Dark Suns*

Art, too, has been using the knowledge of contrast enhancement for a long time. For example, a drawn sun can be greatly brightened in its impression by means of contrast enhancement compared to the image background. An example is shown in Fig. 3.22 on the left.

In reality, the sun has exactly the same brightness as the image background. The illusory brightening is done by a trick similar to the one you learned in the Craik-Cornsweet-O'Brien deception. The sun ends in a sharp transition edge to a dark background ring that blurs to the outside without a sharp transition to the background. Again our perception increases the brightness edges and makes the sun appear brighter and the background darker.

Figure 3.22 on the right shows the opposite case: The sun appears clearly darker than on the left. In reality, however, the two suns are equally bright! The dark impression was achieved by the reverse contrast. The contrast enhancement takes place here at the sharp edge under exactly the opposite sign. Now there is the darker color inside and the lighter one outside! Our perceptual apparatus is outwitted by the continuous brightening of the background from the outside to the inside.

The dark impression created by the contrast enhancement is extended to the entire sun disk.

Fig. 3.22 The two suns are equally bright!

3.3.9 *The Kanizsa-Triangle*

Another trick to brighten up surfaces is the *contour illusion*. How our perception is lead astray by deceptive contours can bee seen in Fig. 3.23 on the left. This picture was created by Gaetano Kanizsa, to whom many charming pictures with contour illusions can be traced (Kanizsa 1955).

Presumably you see at first only a seemingly random arrangement of black circles and lines. These black shapes are in the foreground of the picture. But the figures are arranged in such a way that their imaginary connecting lines can make a completely new sense. As soon as you recognize this, a scene change will take place! Parts of the previous background will jump to the foreground, and vice versa: A triangle now floats in front of a second triangle in the foreground, while the black circles are in the background.

What is most remarkable about this illusion is that the human eye is able to recognize a shape without difficulty, although only parts of its contour are visible. The reason for this is again the strong urge of our perception for the good shape—all parts of the picture now have an ordered structure. All individual picture elements fit together and create a new sense in the overall picture. Our perception apparatus, which is designed for the recognition of edges, is able to supplement or complete edges independently.

In addition to edges and contours that are not present at all, you can also perceive an apparent brightness contrast: The front triangle appears much brighter than its surroundings! In reality, of course, the paper is the same white everywhere. The reason for this brightness deception is probably to be found in certain processes of depth perception. From experience, we are used to the fact that nearby objects are brighter than distant ones. The farther away the objects are, the weaker the light, hazier, and therefore they appear to us less rich in contrast.

This everyday experience probably gives us the feeling that the Kanizsa triangle in the foreground is brighter and more contrasting.

Fig. 3.23 The Kanizsa triangle on the left: A white triangle floats in the foreground! It appears whiter than the white background. Right: A transparent triangle floats in the foreground. The brightness illusion is now removed

3.4 The Inn in the Forest

Our journey leads us from the place on the motorway bridge to a nearby inn—bearing the promising name "The Illus-Inn." It is beautifully situated in the middle of a forest and offers a magnificent panoramic view of the surrounding mountains. The tour group immediately distributes itself to two round tables.

3.4.1 *The Tables*

One table is dark; there are some light beer coasters on it. The other table is light; dark beer coasters lie on it. The coasters on the left are as bright as the table on the right, and vice versa. Viewed from above, the tables with the beer coasters appear as in Fig. 3.24.

The beer coasters seem to be in the foreground because they have a concise shape. Now look at the figures from the point of view of brightness. Surely the figures on the left appear brighter than the table on the right. The beer coasters at the right table also appear darker than the left table. So the figures are perceived with a stronger contrast for similar reasons!

In the meantime, the waiter has brought the drinks and everyone toasts to the fantastic trip. The bus driver is in such a dazzling mood that he bridges the waiting time before dinner with another optical illusion.

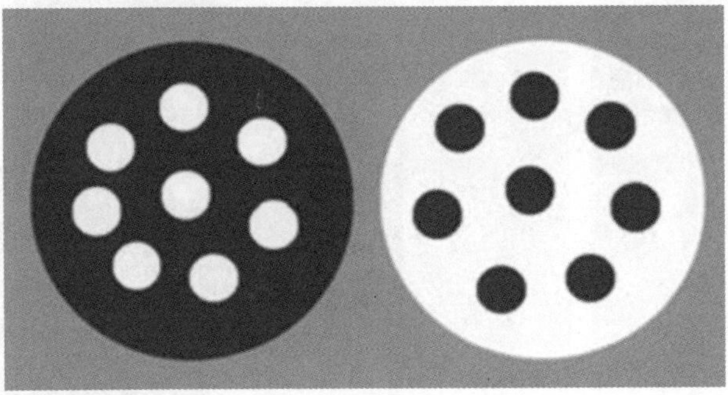

Fig. 3.24 The two reserved tables in the forest inn: Brightness deception due to different figure-background ratios. The beer coasters on the left are as bright as the table on the right, and vice versa. A stronger contrast is assigned to the beer coasters. Therefore the left coasters appear brighter and the right ones darker than they really are!

3.4.2 The Wertheimer-Benary Figure

The bus driver takes a few paper triangles out of his bag and presents the expectant travelers with another example of the strong contrast between the forms in the foreground: the Wertheimer-Benary figure (Fig. 3.25).

This figure goes back to the two Gestalt psychologists Wilhelm Benary (1888–1955) and Max Wertheimer (1880–1943).

The bus driver says: "Compare the two grey triangles, which in reality are equally bright, with each other. Surely the triangle on the upper left seems a lot brighter to you. The reason for this is our perception of form. The left triangle seems to be in front of the black cross. On the other hand, the right triangle appears to be nestled to the cross. This different depth estimation causes the brightness deception. The triangle that appears to be further forward will again have a stronger contrast."

The reason for the different brightening effect of the triangles is the different depth assessment of the figures by our perception. You'll notice it already: this is the same reasoning as in the Kanizsa Triangle.

In Fig. 3.23 on the right you can even see a triangle that floats in the foreground and appears transparent.

The perception of transparency is a very remarkable perceptual quality, which we will now turn to at the end of our journey through figure and form.

3.4.3 The Perception of Transparency

The phenomenon of transparency makes the human eye's perception of form considerably more difficult. Transparency usually implicates a "competition" between different laws of seeing, which does not always lead to a clear solution.

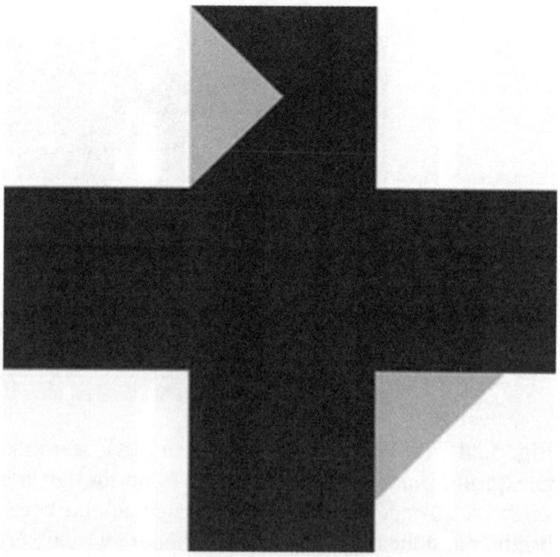

Fig. 3.25 The Wertheimer—Benary figure: The triangle at the top left is brighter than the triangle at the bottom right

Fig. 3.26 The mountain panorama from the forest inn: the glass mountain! In hazy weather, mountains may appear transparent. This is the case when the outline of a smaller mountain at the front fits exactly into the saddle of a mountain farther back. This creates the impression of two new, mutually penetrating, transparent mountain ranges

Have a look at the mountain panorama from the forest inn in Fig. 3.26. Here, some Gestalt laws are "in use"; they cause the senseless perceptual impression of a transparent mountain! In reality there are two mountains: a small one in the foreground and a large one in the background with a saddle in the middle. By chance the boundary lines of the mountains meet exactly at the right place and with the right angle. This creates the impression of two continuous lines!

One line seems to go from top left back to front right down, the other line from left front bottom to right top back. This impression of two continuous lines is only possible in hazy weather, as otherwise the transitions between the mountains are visible due to differences in brightness or contrast.

If the two lines are recognized, our perception desperately looks for a plausible solution for this absurd spatial state, which actually cannot be. After all, it is completely impossible to see two mountains penetrating each other in this way. But our perceptual system finds a brilliant new solution here: the "glass" mountain (described in Metzger 1990). This phenomenon makes the two mountain ranges seem transparent to us!

The impression of transparency thus arises from a mutual overlapping of forms that compete with each other. Which solution our perceptual apparatus finds varies from case to case.

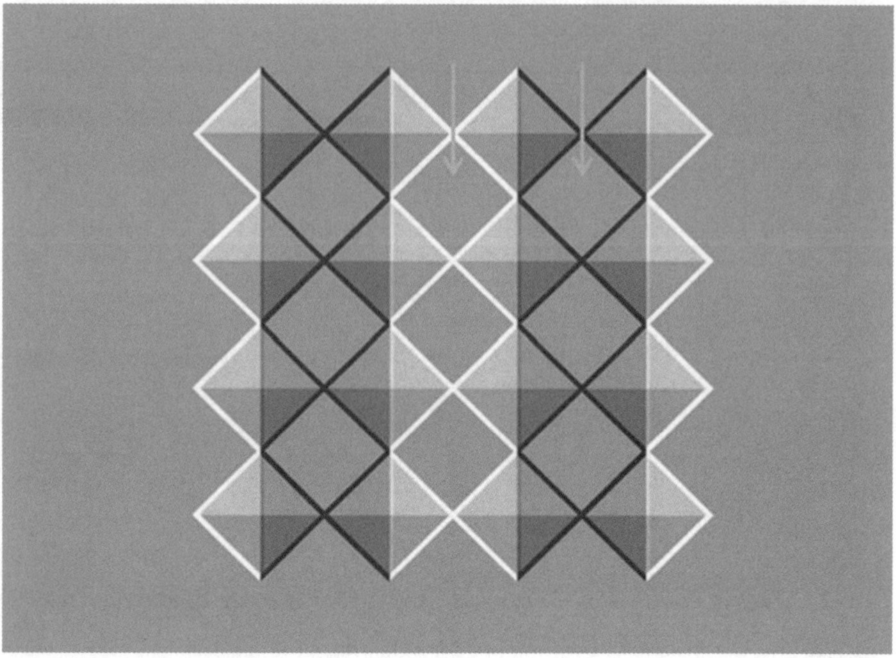

Fig. 3.27 Effect of the sensation of transparency: The two squares marked by arrows are in reality equally bright! (by Edward Adelson)

If, however, our perception takes transparency into account, no guarantee can be given for apparent brightness. This can be seen in Fig. 3.27, which goes back to an idea from Edward Adelson (Adelson 1993).

In this picture, two vertical transparent stripes seem to be located above the background. Our brain immediately draws its conclusions from this presumption of perception: It knows that transparent objects take some brightness away from the background. In order to achieve a uniform, concise background impression, it tries to compensate for this loss of brightness.

It thus "adds" brightness to the background structures superimposed by the transparent stripes. This striving for conciseness is cleverly exploited by the figure of Adelson to create an astounding deception.

Compare the two squares marked with an arrow. Surely you can see a huge difference in brightness, although the two surfaces are indeed exactly the same. You can easily convince yourself of this by covering the surroundings of the two squares with paper!

3.4.4 *White-Deception: Overlapping and Simultaneous Contrast*

The bus driver shows us one last picture before his well-deserved closing time (Fig. 3.28). This is the *White-Illusion* (White 1979), which consists of simple horizontal black stripes and two vertical grey bars. The left vertical grey bar appears much darker than the right one.

This perception is very surprising at first sight, because the brightness contrast enhancement should actually show the opposite effect: The left grey rectangles have mainly black as adjacent color at their edges and the right rectangles have mainly white as neighbor. Due to the effect of brightness contrast enhancement, the left rectangles should appear brighter and the right rectangles darker. That this is obviously not the case is due to the special geometry of the image. Instead of several small grey rectangles we see two vertical bars. The beam on the left seems to lie on a white background—covered by the black horizontal stripes. Conversely, the right grey bar seems to be covered by white stripes and lying on a black background.

The brightness contrast enhancement now simply works against the observed backgrounds, which causes a darkening on the left and a brightening on the right!

With this, the bus driver finishes his explanations, and just in time the dinner arrives for the travelers, but—hard to believe—completely without any further optical illusion.

Fig. 3.28 The White-Illusion

Chapter 4
Fourth Journey: Ambiguous Perception

The world of optical illusions is a fascinating world full of deceptions, idiosyncratic perspectives, dazzling colors, and misguided fantasies. Who sees what and when during the observation of ambiguous images, depends above all on individual previous expertise and personal experiences.

Spontaneous snapshots and witty, sparkling reversal images are the content of this journey. They often seem trivial at first view but all of a sudden gain momentum and life. In addition, our nature is full of ambiguous perceptions that conceal reality—we must therefore first develop the right eye for it. So let us open our senses Germany: "Freiburg, Germany".

4.1 How Did You Find Freiburg?

Today the bus driver has his well-deserved rest day. The group therefore organizes an excursion on its own. Imagine that the starting point of this excursion is in front of a student residence in Freiburg, Germany. At the entrance there is a note with the question:

"How did you find Freiburg?"

"With the map, of course," says one traveler.

"I liked it very much," says someone else at the same time.

Two completely different answers to one and the same question!

Both answers are quite logical and only depend on how you interpret the question. Thus it becomes clear that this journey is about ambiguities.

One interpretation/perception of an ambiguous content is usually possible quite quickly and without any prior information. However, sometimes it can take quite some time until one recognizes possible further alternatives for the interpretation of the content—if one notices them at all. The longer it takes, the more beautiful is the aha experience afterward.

T. Ditzinger, *Illusions of Seeing*, https://doi.org/10.1007/978-3-030-63635-7_4

You can try this out, for example, on the next piece of paper on the door of the residence, which is inscribed, "I fell in love with Freiburg". This sentence again enables two different interpretations of content. All right?

By the way, most of the jokes in which an ambiguous word or phrase occurs are based on this surprising aha experience in understanding the second possibility of interpretation. You can see this from Fig. 4.1, which is displayed on the entrance door of the dormitory.

"It escapes me" says one participant.

Fig. 4.1 The picture at the entrance door

"Escaping is easy," a passing student answers, laughing, and runs away.

"I'm not going with you," the traveler says and shakes his head.

"Then stay right where you are," says the student with a mischievous smile, and opens the door to the dorm, snorting. And with the beat of the door *leaf* he moves his nasal *leaf*, takes a seat at the *leaf* (piano), and while he watches the first *leaf* of autumn falling down he starts playing a wonderful song which he reads from a *leaf*. The title of the song? "Fallen Leaves" from Billy Talent!

After a short pause for thought we open the door together with our friend from the travel group, who is also laughing in the meantime, and look around in the mysterious dormitory.

The ambiguities presented so far all originate from the area of language. In the following, however, we will deal with ambiguities in vision. A first glance at the corridor of the dorm gives us a clue, because the walls are covered with pictures all over.

4.2 The Rubin-Cup

Look at Fig. 4.2 for a few seconds. After you have fairly quickly gained a clear interpretation of the picture, your perception will automatically skip over to another, alternative interpretation. So you can interpret Fig. 4.2 on the one hand as a white vase of flowers and on the other hand as two black faces.

Images of this kind, called *ambivalent images*, always make several interpretations possible. They are thus the visual counterparts to the linguistic ambiguities encountered above. Figure 4.2 is a variant of the famous *Rubin-cup*, the original of

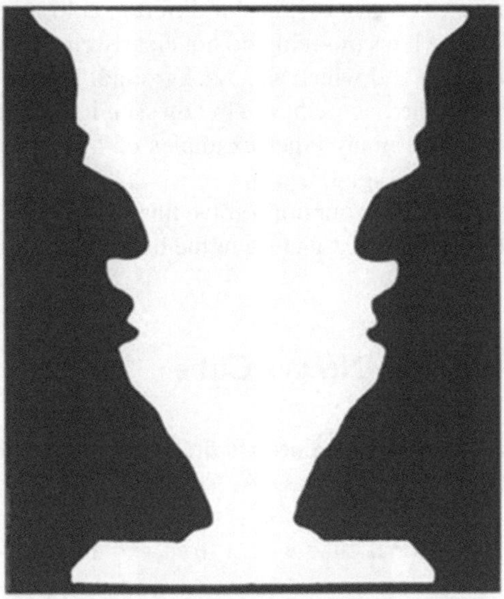

Fig. 4.2 The Rubin-Cup: vase or faces?

Fig. 4.3 Figure background pictures. *Left*: Signet of the 1990 World Chess Championship. *Right*: Saxophone player or face?

which dates from 1921 and is named after its inventor; it is counted among the group of *figure-background images*.

These figure-background pictures are flat drawings in which the viewer can recognize the various surfaces as both foreground and background. Thus our system of perception gets into a serious, even insolvable conflict. The reason: the different viewing possibilities are normally mutually exclusive. The figure-background images consist of areas of different coloration and brightness.

The laws of seeing do not clearly clarify which part of the surface is perceived as a figure and which as a background. Rather, the laws of seeing even compete with each other, as we have already seen in the second journey. Beside the vase of Rubin there are many other examples of figure-background pictures, some of which are shown in Fig. 4.3.

How does our brain solve this supposedly unsolvable conflict? To clarify this, we look at the next picture in the hall of the dorm.

4.3 The Necker Cube

Figure 4.4 is the line grid drawing of a three-dimensional cube which is named after the mathematician L. A. Necker (Necker 1832).

You probably don't suspect anything special behind this drawing: a three-dimensional cube with a front and a back—nothing more. *Which surface is in the*

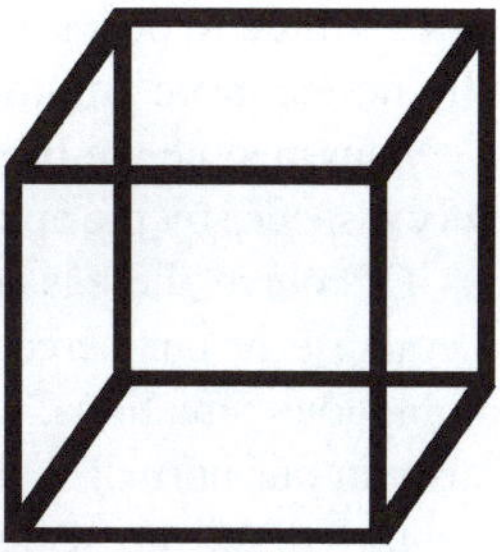

Fig. 4.4 The Necker cube: Take some time for this picture: Suddenly it starts jumping back and forth like wild in space!

Fig. 4.5 Two areas that could be in the foreground

foreground? is the question that brings us closer to the secret of Fig. 4.4. So which surface is in front?

"The surface that ends at the front left," a traveler means.

"No, the surface at the top right," another asserts strongly.

To clarify their claims, the two hatched in each case the surface of the cube, which they believe is in the foreground, in different colors (Fig. 4.5).

After the two of them have been insisting on their opinion for a while and have continued to excitedly stare at the cube, suddenly and almost simultaneously a change in their behavior occurs.

"Now I suddenly see it in the other perspective," says the one of the travelers.

"Actually, with me it jumped all of a sudden as well", the other replies, completely surprised.

This impressive phenomenon of the turning spatial cube is called the *Necker cube*. After its discoverer.

Take the time to look at the Necker cube a little longer—it's worth it! Pretty soon you will very probably become aware of a spatial alternative of perception. Now just keep looking at the cube! The jump over of the cube then comes on its own and completely unexpectedly! Let yourself be enchanted by this trick in which you are both magician and audience at the same time.

Once again, our brain has solved an apparently unsolvable problem in an initially completely surprising way. A simple, still image like the Necker cube sets our perceptual apparatus in motion!

Continue to observe the figure after the first jump of perspective. You will gradually notice more and more stable oscillations between the two spatial alternatives.

Many people are frightened when first experiencing this phenomenon and become very insecure by the apparent collapse of their supposedly so stable perceptual apparatus. Of course, there is no reason at all to be frightened. In fact, we are not experiencing a collapse of our perception, but rather another fantastic trick of nature to deal with ambiguous situations. This process of closing compromises occurs almost continuously in our normal world. In our everyday life everything is in motion, new environmental stimuli are constantly pouring in on us. Thus, our perceptual apparatus is constantly faced with the task of mediating between these stimuli, filtering out the most interesting and, if necessary, alternating between several alternatives. More generally: Our entire environment can be seen as one fascinating ambiguous task of perception.

The fascination of looking at ambivalent images such as the Necker cube is based on the fact that they show us the temporality of perception in a particularly drastic way. Nobody expects at the first viewing of such a picture that it will "come to life" and—as if moved by a ghost—seems to jump back and forth!

The claim of Pöppel (2001) that the human brain can only ever have *one* perceptual content in its consciousness is confirmed by these perceptual processes when viewing an ambivalent image such as the Necker cube. As soon as we become aware of a new perspective, the old one disappears completely.

All these ambivalent images are constructed in such a way that they leave at least two choices open to us when we look at them. With the Necker cube this is achieved by the fact that no spatial representation of the cube is preferred: The two squares, which alternatingly form the foreground and the background, are completely identical in size!

Our perception has discovered the perfect compromise and simply alternates between the different possibilities.

There's even another way to look at the Necker Cube—without any spatiality! This possibility can be easily understood by looking at the following series of images (Fig. 4.6) of the spatially rotated Necker cubes from left to right.

While in the case of the cube on the left side the already known oscillations between the two spatial orientations take place, in the cubes in the middle and on the right a planar component comes more and more into play. With the right cube, this

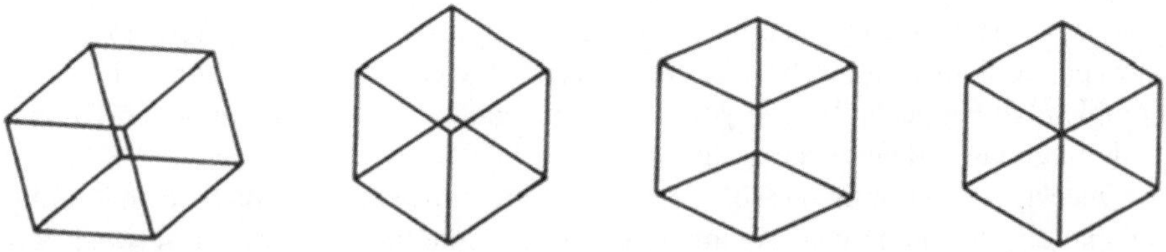

Fig. 4.6 Spatially rotated Necker cubes according to Kopfermann: From left to right, a planar perception is increasingly possible

alternative even dominates so strongly that spatial perception is very limited. The oscillations of perception are strongly hindered here!

As diverse and fascinating as the ambivalent patterns themselves are, so diverse are their designs and forms. So let's keep looking around on the wall of the dormitory. There's a whole lot of other line drawings on it that exploit the effect of the ambiguity of space.

4.4 Ambivalence of Perspective

With perspective ambivalence, the spatial orientation of three-dimensional objects is not clearly defined, so that several three-dimensional views are equally perceptible. As with the Necker cube, oscillations of perception occur again after some observation time. Some examples are shown in Fig. 4.7.

In the meantime, our tour group has arrived at another dispute case, shown in Fig. 4.8.

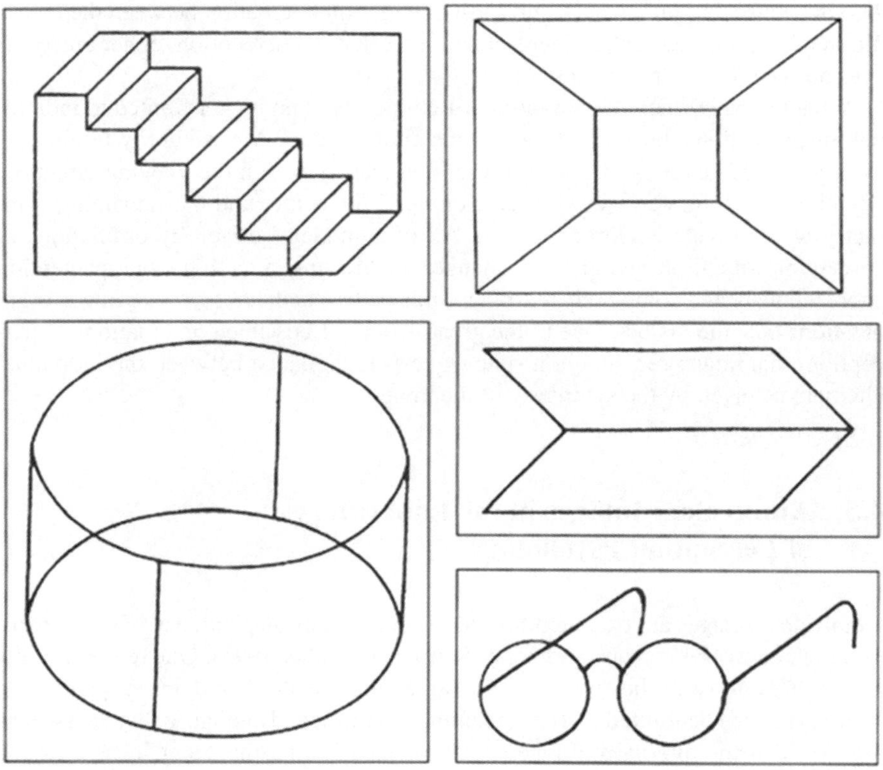

Fig. 4.7 Examples of perspective ambivalence: All these objects can be seen in two different spatial orientations!

Fig. 4.8 One or four dice? One or twenty-four eyes?

Here a different number of dice can be perceived: either one or four dice. Our perceptual system in its need of harmony does not "invent" two and a half cubes as a compromise, but jumps back and forth in temporal alternation between the recognition of one and four cubes. Depending on the state of perception, either one or 24 eyes are visible—a dream trick for every cheater!

What happens when the possibility of ambiguity is no longer limited to individual image sections, but extends across the entire image? A particularly fascinating example of this case can be seen in Fig. 4.9. Depending on the viewing direction, attention and point of view, you see a circle in the center and the remaining symmetry of the image is aligned with it. All of a sudden—driven by oscillations of perception, attention swings and changes in viewing direction—our perception spontaneously and completely rearranges this pattern and other circular image organizations become visible. Due to the great variety of possible combinations of perception, this image remains in a state of perpetual change between the individual alternatives given by the symmetry of the image.

4.5 Ambivalent Images in the Laboratory of Perceptual Psychology

Ambivalent images are excellent for measuring the time dependence of our perception (a good overview you can find at (Kruse and Stadler 1995). One reason for this is that, in contrast to the "normal" moving environment, the number of perception alternatives can be limited to a small number. A further advantage is that the switching between the alternatives takes place so clearly that the point in time of the change can be unambiguously named by each observer.

The test normally takes place in such a way that the test subjects have to press different buttons according to their perceived alternative. With these simple

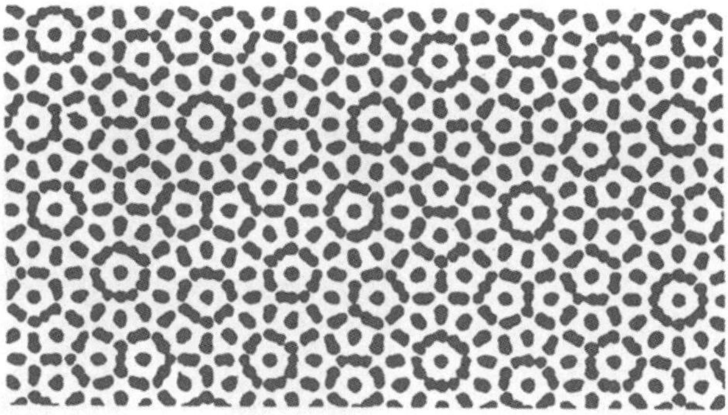

Fig. 4.9 Large-scale oscillations of perception. (With the kind permission of Michael Stadler)

measuring methods, one found out quite amazing things. For example, it was shown that the oscillations of perception can be delayed by free will, but not stopped at all. You can easily check this if you look at the Necker cube in Fig. 4.4 with the specification that you want to hold on to the spatial perspective you have seen. Inevitably you will soon have to capitulate to your own perceptual system and have to allow the change of perspective.

4.5.1 Measurement Method

If you look at Fig. 4.4 for a little longer, you will surely notice that it takes some time before the oscillations really "warm up." Only after the relatively long time of about 3 min do really stable oscillations of perception occur. Following this familiarization phase, the actual laboratory test begins. It was found that on average there is a change of perception every 3 s. However, there are greater differences from person to person in the average duration of perception of a perspective. This personal property can be seen well when looking at Fig. 4.10. This is a Maltese cross with eight cross-segments. Either a black cross on a white background or a white cross on a black background can be seen. From person to person there are significant differences in the speed of the change of alternatives.

4.5.2 The Oscillation Speed as a "Fingerprint of the Psyche"

This difference in the speed of the change of perception was investigated in a laboratory experiment. It has been proven that each person has their own unique oscillation speed.

For people with a high speed of change, there are only very small deviations in the duration of perception. In other words, such "fast" people have a very constant

Fig. 4.10 A Maltese cross with eight wings

rhythm of change of perception. In contrast, people with slow changing speeds have very high fluctuations in the lengths of their perceptual durations!

The oscillation frequency thus represents a characteristic quantity for every human being. Under the same conditions, the duration of perception is a perfectly reproducible quantity for each test subject.

To date, however, it has not been possible to identify any special characteristics of the test subjects as the clear cause of this variable. For example, the influence of the intelligence quotient, introversion or extroversion, brain damage, or the use of drugs were considered.

Just as with a fingerprint, no conclusions can be drawn about personal characteristics such as height, hair and eye color, or mental condition. However, as with fingerprints, the person tested can be identified by the oscillation speed. The oscillation speed thus has the meaning of a "fingerprint of the psyche."

4.5.3 Pictures with Different Weighting of the Alternatives

Only in the rarest of cases do the alternatives of the perception of an ambiguous image come to bear equally strongly. In order to produce such equivalent images in a targeted manner, the laws of vision involved must be observed and applied very carefully.

A good example of this is the influence of the law of proximity. Have another look at the picture "Vase or face" by Rubin (Figs. 4.2 and 4.11).

The distance between the black faces in Fig. 4.11 increases from left to right. Thus the weighting is shifted more and more from the alternative "Vase" to the alternative "Faces"! The simple reason for this effect is provided by the grouping law of proximity, which you already got to know during our first journey. In the left

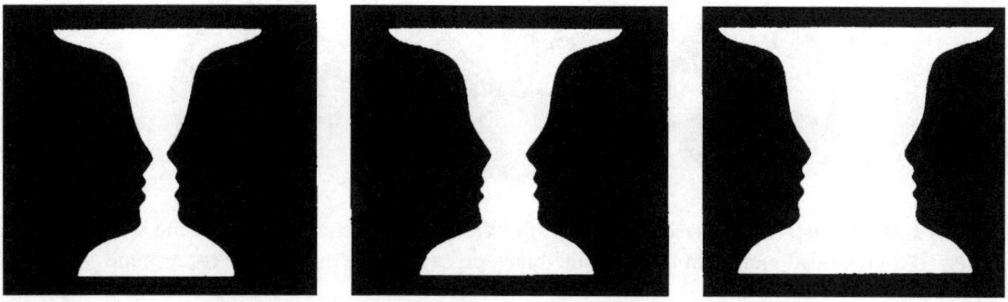

Fig. 4.11 "Vase or Faces?" with different distances between the faces

picture the white vase is emphasized by the proximity of its outlines in contrast to the right picture and becomes thereby the dominating figure.

Even with ambiguous images with different weighting of the alternatives, oscillations of perception take place! However, the duration of perception of the alternatives varies: The more an alternative is preferred, the longer the average perception time of this alternative lasts. The average duration of perception of the weaker alternative is about 3 s. The duration of perception is a good quantitative measure for the weighting of the individual images. The duration of perception is therefore an ideal indicator for measuring the influence of the laws of vision involved in the weighting of alternatives!

We can understand this with the example of the Maltese cross with eight wings. Figure 4.12 illustrates the influence of the different orientation of the pattern on perception. From left to right the Maltese cross is turned step by step around its axis.

Probably you see the white cross in the foreground in the first picture—the explanation for this is the already discussed vertical illusion. The law of conciseness prefers the vertical/horizontal arrangement.

For this reason, conversely, the black cross is strongly preferred in the third partial image, while orientation has no influence on the perception process in the second partial image—this was confirmed by laboratory experiments carried out by Oyama (1960).

At the Maltese cross one can observe further amazing perception effects. Consider, for example, the series of three Maltese crosses shown in Fig. 4.13.

The base area of the black cross increases strongly from left to right. In the left picture you may see a black cross on a white background. However, with the proportion of black increasing from left to right, this impression is lost. In the right picture the perception of the white cross is clearly preferred.

This observation was proven experimentally by Künnapas (1957). By measuring the duration of the individual periods of perception, he accurately measured the increasing dominance of the alternative white cross from left to right.

To explain this phenomenon we can again refer to the law of proximity. A small, closed surface appears much more like a figure, i.e., lying in the foreground, than a larger surface surrounding it. Therefore, the black cross is more visible in the left figure than in the right figure and the white cross is clearly visible in the right figures.

Fig. 4.12 Three twisted Maltese crosses. In the first partial picture at the left the white cross is preferred, in the third picture at the right the black cross appears to be in the foreground

Fig. 4.13 Three Maltese crosses with different base areas: The white cross appears increasingly to be in the foreground from left to right

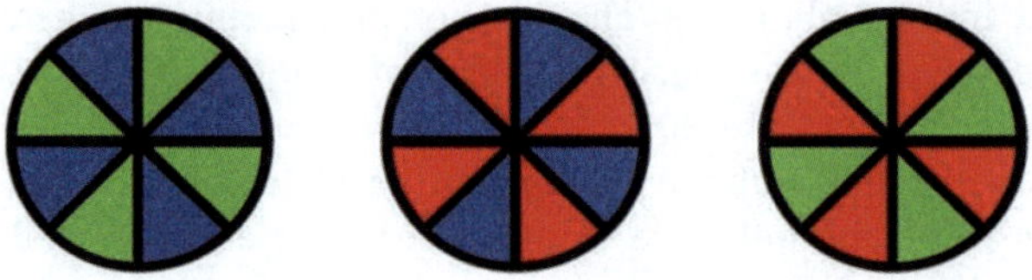

Fig. 4.14 Maltese crosses in different colors: The red cross is on average perceived before the green and the blue one

What do you see when you look at the colored crosses in Fig. 4.14? Surely you see a certain color preferred, whose cross then appears to be in the foreground.

Studies have shown that every human being has his own personal "color sequence." With the help of a large number of test subjects, an average sequence of colors was determined in the laboratory by measuring the oscillation times. Oyama found that most people see the red cross in the foreground, followed by the green one, whereas the blue one appears most often in the background.

4.6 Young Man or Father-in-Law?

Now look together with the travel group at the next picture in the hallway of the strange dormitory (Fig. 4.15).

This time all participants agree that they see a man. That's something, after all! But how old do you think this very special man is? He is so special because the assessment of his age by different people can certainly differ by 50 years—which is

Fig. 4.15 Young man or Father-in-law? (After Botwinick (1959))

Fig. 4.16 Two phantom drawings

still remarkable even in the age of cosmetic surgery! For clarification, see the two phantom drawings in Fig. 4.16.

The differences are now clearly visible: the young man looks back to the right and has blond hair, while the old man looks down to the right and has black hair. In addition, the two heads differ in many ambiguous details: the chin of the young man is the nose of the old man, the hair of one is the eye of the other, and the collar of one is the mouth of the other. These ambiguous details give this picture its very special charm. With a little patience you can observe the continuous changes in your own perception.

You can make the change of perception much easier with a trick: Since every detail of the picture is ambiguous, the viewer only has to choose a certain detail that prefers the desired alternative. If, for example, you cover the upper half in Fig. 4.16, the perception of "old man" is normally somewhat preferred, whereas the young man can be seen more easily when covering the lower half of the image.

4.7 Young Girl or Mother-in-Law?

You can reconfirm this method by looking at the similarly constructed famous picture from Hill shown in Fig. 4.17. It is the female counterpart to the previous picture.

Again, the viewer can gain a completely different impression of the woman depicted. Either a young woman looking back to the left or an old woman looking forward to the left can be perceived. For example, the necklace of the young woman becomes the mouth of the old woman. The decision for a point of view is made immediately, so to speak at first glance. But how does this decision come about? Can a conclusion be drawn from this assessment at first glance as to the personality or daily form of the observer? And can the timeless question "Am I a good lover?" finally be scientifically evaluated?

This famous picture has been rediscovered over the years. The original was probably drawn by an unknown German artist from 1888 (published on a postcard). See Fig. 4.17.

4.8 How Does Our Brain Make Decisions?

The approach of our brain in recognizing such images is that of a spontaneous break of symmetry. The two existing possibilities of perception represent a symmetrical, equal state at least in our subconsciousness. Here, our brain proceeds according to the same decision-making scheme that is known to us from completely different areas of human life.

There are plenty of cases in which two or more alternatives of action are given, which are exactly the same or comparably good. Decisions have to be made all the time: Do I put on the white or blue socks, do I have another cup of coffee or not, do I finish reading this page of the book or not? And so on. Fortunately, for most people, decision-making in such small problems is largely automated by everyday life. Nevertheless, there are people who, in such cases of doubt, always seem to be stuck between exactly symmetrical alternatives and can never decide for or against something.

The reason why some people cope very well with such choices and some very poorly is our respective internal value systems. The decision neurotic often artificially assesses the possible alternatives in the same weight out of fear of a wrong decision and thereby delays a decision as long as possible. At the same time, however, he suffers greatly from this situation. Another kind of person, on the other hand, finds it very easy to make everyday decisions. These people enjoy the process of decision-making as a special freedom of life.

No matter how long a state of symmetry is maintained, at some point a decision is made and symmetry is broken. We can also observe similar things in important life decisions: For example, a woman wavers between two men who have different merits—which one should she choose? Or you get an offer for a new job in another place. Are you accepting or refusing this job? For both possibilities there are advantages as well as disadvantages that have to be weighed against each other. The decision is made by our brain in a very simple way—you can easily understand that with the help of Fig. 4.17. Here, too, several picture details speak for the alternative young woman and several for the alternative old woman. Which alternative is

Fig. 4.17 Young girl or mother-in-law? *Above*: the most famous picture after Hill (1915). *Below:* the presumed original from 1888 (anonymous German artist)

ultimately perceived is therefore almost pure chance. A tiny trigger at the right moment leads to the decision: Which part of the picture do we look at first? Did we maybe just see a young woman walk by outside? What are the current lighting conditions like? And so on. As you can see, a small cause can have a big effect if it comes at the right time. Before and after the decision is made, however, these environmental fluctuations are usually quite unimportant.

4.9 Synergetics

The German physicist Hermann Haken calls this mechanism *symmetry breaking by critical fluctuations*. These technical terms originate from the science of *Synergetics*, which he founded (Haken 1977). It uses mathematical methods to explain the self-organization of many-particle systems in the animate and inanimate nature. It turned out that the basic mechanisms of these self-organization processes always follow the same scheme, whether in physical or biological systems, in psychology, economics, or sociology.

The constantly existing fluctuations have no influence on the stability of the system in the normal state. However, when the system is in a symmetrical state, fluctuations can become critical and temporarily gain control over the system. This is the case until the system is in a new stable state. A *phase transition* has thus taken place.

A simple model for this process is a sphere located exactly in the middle between two symmetrical valleys (Fig. 4.18). In this unstable state the fluctuations become critical: a slight breeze in one of the two directions, for example, is sufficient to achieve the phase transition of the sphere into one of the two valleys. Once it arrives in the valley, the fluctuations are completely uncritical again, i.e., even a strong gust of wind hardly leads to a change in condition.

This model also makes it possible to understand the processes of human decision-making. For example, the question of which man a woman chooses depends on critical fluctuations, as does the question of the right place to work.

The moment the decision situation is exactly 50 to 50, the outside world takes on great importance for a short, important moment. This can be a phone call or a gift at

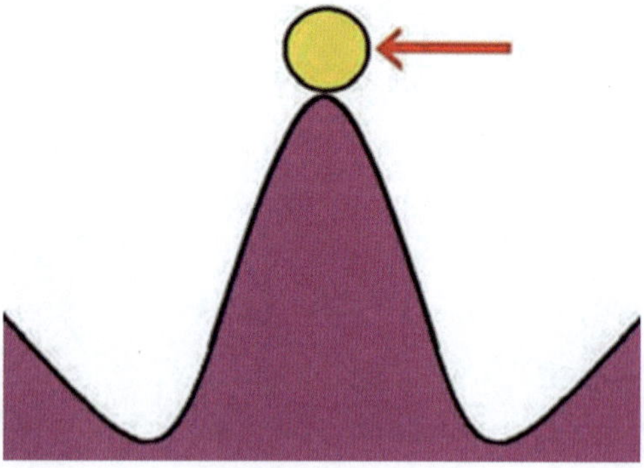

Fig. 4.18 Example of a symmetry breaking due to critical fluctuations: A small breeze from one side is enough to decide whether the ball rolls into the left or right valley

exactly the right time, but it can also be exactly the opposite. Often, however, as in the recognition of the perceptual alternatives of an ambivalent image, external influences on the subconscious play the main role: perhaps a car passes by that has the same color as the car of one of the two men; one hears the laughter of a passerby that reminds one of someone; one finds a coin on the street, and so on. The main thing is that the timing is right!

4.10 The Prejudice

Let us turn again to Fig. 4.17. Perceptual psychological experiments by Leeper (Leeper 1935) with numerous test subjects revealed the following statistical distribution: 60% first recognized the young woman, 40% the old woman—the mother-in-law. Measurements showed that the randomly fixed image section was the main factor of influence for unbiased persons.

However, if the persons were previously prepared accordingly, the other factors became more important. For example, if the subjects were shown an unambiguous image of a young woman at the beginning, in the subsequent experiment with Fig. 4.17 all subjects recognized the young woman first. Conversely, if an image of an old woman was shown first, 94% of the test subjects opted for the alternative old woman when subsequently viewing the ambivalent image.

A good example of bias through prior knowledge can be seen in Fig. 4.19. In the figure on the left you can quickly see a “13” or a “B”. As can be seen in the figure on the right, this ambiguous figure quickly becomes unambiguous through an added context. If one looks at the figure from top to bottom, the “13” is clearly recognized; if one looks from left to right, however, the result is clearly a “B”.

Context dependence also exists under water and in Australia. You can convince yourself of this with the help of Fig. 4.20.

The focus is on the white ambiguous outline, which is shaped by the respective background pattern. The background “water” in the left sub-picture creates the perception “dolphin.” On the other hand, the implied plains background transforms the dolphin into a kangaroo.

Fig. 4.19 “13” or “B”. (Courtesy of Michael Stadler)

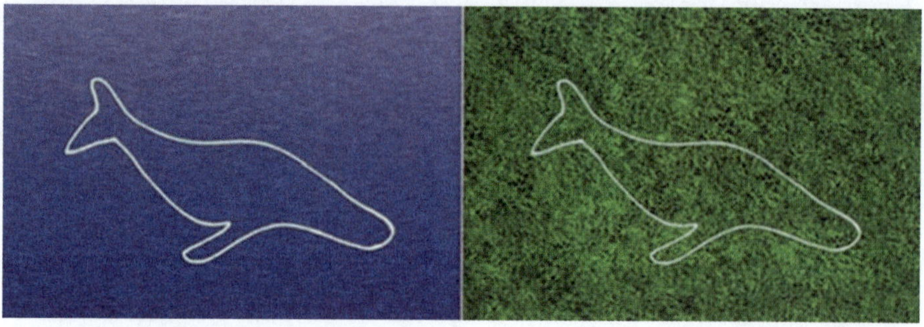

Fig. 4.20 Dolphin or Kangaroo?

Fig. 4.21 Rabbit or Duck? (Jastrow 1900)

However, such a bias often does not even have to be artificially created, but can be rooted in us from birth. For clarification, look at the next ambivalent image in Fig. 4.21. Hartmann and Heiß used a similar model as in Fig. 4.21 in a laboratory experiment in 1962 with a dominant duck percept. Out of 26 test subjects, 23 recognized a duck and only three a rabbit.

The remarkable thing about this attempt was that two of these three people were called "Haas," which is homonymic to the German translation of rabbit ("Hase")! Understandably, these people are strongly biased by their previous lives, and their subconscious decides accordingly!

If the respective stimulus pattern contains only little information and thus leaves enough room for interpretation, our subconscious mind takes on the leading role in perception. This is the case, for example, with the detection of color spots in the Rohrschach test of psychologists, of single clouds passing by or even of simple pencil lines. A nice example of this can be seen in Fig. 4.22.

When looking at this line, you—especially the men—probably first recognize the figure of a woman's body, according to Sigmund Freud, while at some point a face probably appears as a second possibility of perception.

Fig. 4.22 A simple line (or?). (With the kind permission of Michael Stadler)

Fig. 4.23 Reverse images. *Left*: Comic drawing from the series "Upside-down." *Right*: dog or cat?

4.11 Reverse Images

Figure 4.23 on the left shows a bird holding a small human in its beak. This is a completely normal, unambiguous picture with no further ambiguous meaning.

Really? If you look at the picture from a different perspective, there is suddenly a fantastic change in the organization of the picture elements. Please turn the book upside down and look at the picture anew!

The picture comes from a comic series by Gustave Verbeek, which was published around 1900 in the *Sunday New York Herald*. Verbeek actually wanted to draw 12 pictures in his story, but for reasons of space he had to limit himself to the usual cartoon length of six pictures at the time. That's why he designed all the pictures in such a way that they turned upside down to become another picture of the plot. In the first six pictures, his heroine, named "Little Lady Lovekins," finds herself in a dangerous situation, which finally dissolves into pleasure after turning the picture around. The hero, named "Old Man Muffaroo," comes to her aid and saves her. If you look at Fig. 4.23 normally, you will see Little Lady Lovekins in the beak of a giant bird. Now turn the picture upside down! The scenery is completely transformed. Now you can see the hero sitting in a boat being followed by a giant fish.

Figure 4.23 on the right, which shows a cat, works with the same trick. As soon as you turn the picture upside down, you will see a dog with ears raised!

These images all belong to the class of *semantically ambivalent images*. Further examples of this most impressive category of ambivalent patterns can be seen in Fig. 4.24.

How does this semantic ambiguity arise? The reference and centering relationships of the image details are not clearly defined in semantically ambivalent patterns.

These are two completely different alternatives of perception in terms of content, which are visually so similar that both can be perceived. Semantically ambivalent images are particularly appealing because this difference in content can refer to any image detail—you can easily understand this using the examples in Fig. 4.24: A total of five animals and three people can be seen on it.

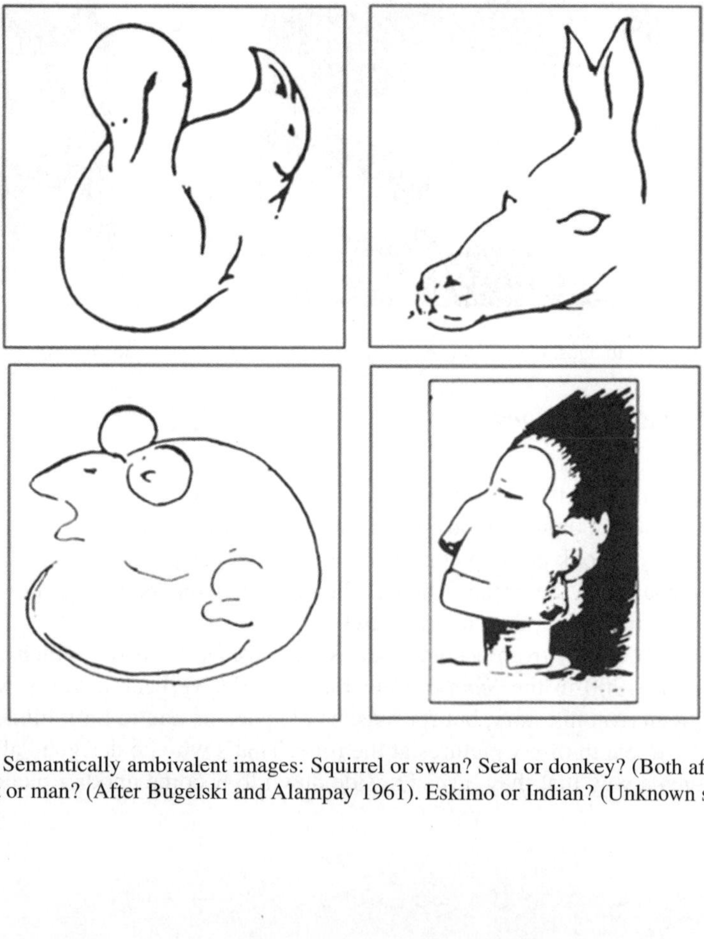

Fig. 4.24 Semantically ambivalent images: Squirrel or swan? Seal or donkey? (Both after Fisher 1968). Rat or man? (After Bugelski and Alampay 1961). Eskimo or Indian? (Unknown source)

Fig. 4.25 The mountain pictured is called Ixtaccíhuatl (5286 m, Mexico), which means as much as "The Sleeping Woman." Why, you'll see for yourself after a while

Even landscapes can have an ambiguous appearance. You can see this for yourself in Fig. 4.25.

In speech perception, by the way, there is an interesting analogy to the semantically ambivalent patterns, the *Verbal Transformation Effect*. To illustrate this, please repeat the word "life" several times in quick succession—i.e., "life, life, life, life" and so on. Until *life* turns into *fly*!

4.12 Morphing

Nowadays there is a multitude of computer programs that can gradually transform not only old men into young men or life into flies, but even the most diverse images into each other. Programs with these properties trade under the collective term *morphing*. The first programs, which were able to merge different line drawings into each other, have existed since 1973 (Wilson).

During morphing, the start image is split into small image elements that are assigned to the final image manually. The program then gradually calculates the transitions between these extreme states. If you put these intermediate stages together, you get whole animation sequences whose fascination is stunning.

With this "magic trick" you can at least visually transform any object into any other object. It's no wonder that advertising in particular takes advantage of this surprise game!

Fig. 4.26 Morphing: A molehill turns into a mountain

In the meantime, morphing algorithms have become so effective that practically anyone at home can calculate their own transformation sequences with the help of a computer or a smartphone app. You can, for example, transform your boss into anything you want. The program also turns a molehill into a mountain, as seen in Fig. 4.26—a trick of transformation that only man seemed to be capable of until now.

4.13 Hysteresis in Perception

As you can see in Fig. 4.27, there were even such gradual (hand-drawn) transitions between different perceptual alternatives as early as 1967. It becomes clear that even man and woman can have things in common—in case you ever doubted it.

Starting from the head of the man in the top left corner, the details of the picture are changed step by step in the direction of the young woman, who is clearly recognizable as such in the bottom right corner. The drawing file at the top right has the most balanced proportions of both alternatives.

With this sequence of pictures another impressive self-experiment can be carried out. To do this, look at the sequence of pictures from top left to bottom right for a few seconds one after the other! Follow exactly how long you recognize the man's

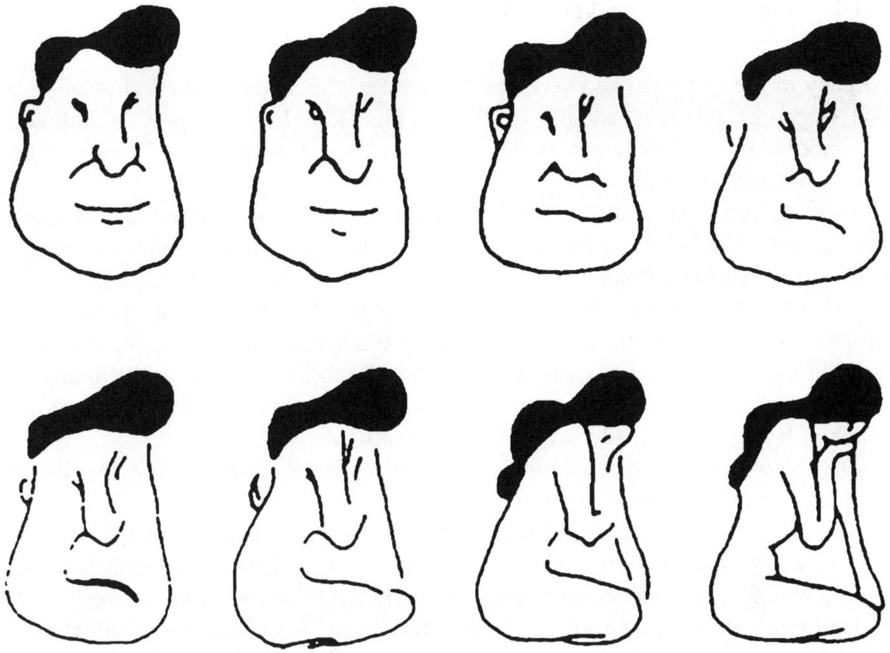

Fig. 4.27 Hysteresis in Perception: First look at the drawing files step by step from top left to bottom right and then in the opposite direction. Depending on the viewing direction, the change in perception takes place at a clearly different point. (According to Fisher 1967)

head! Surely you are hesitant only in the lower line. Normally, when looking at the last pictures, the transition to the perception of "young woman" takes place.

Now carry out the test in the opposite direction! Starting from the picture at the bottom right, you will certainly only experience the transition of perception in the direction of "man" somewhere in the upper line.

This delay in the alternation of the two alternatives is called *hysteresis in perception*. The explanation for this phenomenon is obvious when we recall our knowledge about the biases of test subjects. The picture currently being viewed is always a pre-preparation of the next picture—with the only difference that it is a whole sequence of individual pictures.

In the meantime we have seen all the pictures hanging in the hall of the dorm—except for a sign. It hangs on an old oak door through which we enter a large hall. The sign says:

Fantastic Art Gallery: Special exhibition of ambiguous images, open throughout.

4.14 The Fantastic Art Gallery

Already in the sixteenth and seventeenth centuries many painters created ambiguous works of art. An example of this is Fig. 4.28 by Giuseppe Arcimboldo. Arcimboldo (1527–1593) was a court painter in Prague and decorated, among other things, Milan Cathedral.

Here the drawn summer fruits are arranged in such a way that the face of a man can be seen. Fig. 4.28 is part of Arcimboldo's famous seasonal cycles, in which heads are composed of realistically painted fruits, flowers, and vegetables.

Salvador Dali, one of the most fascinating painters of the last century, also adopted this method of drawing ambiguous pictures. Around 1933 he became interested in ambivalent figures. It is assumed that the inspiration for Dali's appealing multivalent pictures came from special effects and reflections of sunlight on the beach of Rosas on the Costa Brava, where the artist lived.

Dali himself speaks of "seemingly quite normal pictures inspired by the frozen, very small riddle of certain snapshots."

These ambiguous images of his Rosas series perfectly combine image and illusion. Convoy Maddox writes that "they capture the rushed hunch of secret activities that lie halfway between actual reality and the realm of unconscious desire."

Dali made the phenomenon of multivalent images world-famous with unprecedented perfection and beauty by creating a new art form with them. Known examples of these pictures of Dali are: "Phantom Cart" (1933), "Metamorphosis of Narcissus" (1937), "Spain" (1936–1938), "Slave Market with the Disappearing

Fig. 4.28 Giuseppe Arcimboldo: Summer (1563)

Bust of Voltaire" (1940), and the "Apparition of Face and Fruit Dish on a Beach" (1938). In this remarkable picture, there are not only some normal ambiguous images but also a lot of them. In the picture a dog can be seen whose head is part of a beach and whose back consists of fruits. The belly of the dog may also be perceived as part of a white bowl. The foot of this bowl can again be perceived as the back of a nun.

Other modern examples of ambiguous pictures can be found in the works of Maurits C. Escher, René Magritte, Viktor Vasareli, Franco Grignani, and above all Sandro Del-Prete.

In his works, Escher mainly used the ambivalence of figure and background. The different figures, which can be perceived alternately, are surfaces that are constructed in such a way that they fit exactly into each other.

Magritte used a special technique to arrange various symbols and images in his works. The mystical ambiguity of his works arises from the ingenious, content-related connections of these symbols. But also the ambiguity of space was cleverly exploited by Magritte. Magritte himself believes that these different points of view are exactly the possibilities to see the whole world: "We see the world outside of ourselves and at the same time we only have an image of it in us."

Sandro Del-Prete has concentrated entirely on the art of semantically ambivalent images, creating a fascinating collection of ambivalent images.

A wonderful picture with the title "A New Day" was created by the Swiss artist Robert Fischer (Fig. 4.29). This picture is currently being distributed as a postcard

Fig. 4.29 Robert Fischer: A New Day (1991/92)

Fig. 4.30 Wolfram Nagel Cat and Bird (1997), Lion and Face (1997)

and can be used, for example, to announce a birth—although at first no baby is visible! Only after some time does the second alternative of perception open up.

The graphic artist Wolfram Nagel uses the art form of semantically ambivalent images in his works, preferably drawing with pencil. Two examples of this can be seen in Fig. 4.30. Surely you can recognize a cat's head right away. But there is something else to discover: Every cat would be happy if this other animal would come so close to it.

Also interesting is the genesis of this picture. The artist has long experience in drawing birds. This helped him to a "second glance" when looking at a cat by chance. From this biased basic idea, the present picture was created after gradual approximation of the contours and image details of the two animals.

If you haven't found anything yet, cover the left half of the cat and slowly turn the picture clockwise. Now, at the least, the bird's head should appear in front of you!

In the right picture you can see either a lion or a man. The ambiguity of even small image details can be easily observed in this example. Thus the nose of the lion functions at the same time as the chin of the man.

The ambiguous artworks considered so far consist of the combination of two or more perceptual alternatives in one single image.

The gradual approach or transformation of the alternatives has so far played a role at most in the genesis of a painting in the studio. It is precisely this process of transformation that is the special stylistic device of Fig. 4.31 by Silke Haarer, entitled "Metamorphosis." It depicts the transformation of not quite everyday transformations of quite everyday things into each other, which at first glance are not connected to each other in any way.

From left to right, a wrench turns into a fish, a fir branch with a crystal ball into a bony fish, and a candle into a toilet with a toilet roll and old toilet water flush.

Like in a fantastic dream, things gradually change shape, orientation, or color—or everything at the same time. For example, the red crystal ball changes its color to blue, the orientation of the fir needles is different from the orientation of the bones, the shape of the left end of the wrench changes in the head of the fish. The

Fig. 4.31 Silke Haarer: Metamorphosis, 1994

transformation also takes place by repeatedly taking away and adding entire parts of the image. For example, the fins of the fish or the flushing of the toilet are added, while the rays of the candle are removed. The two lower rows show the gradual transformation of an apple. In contrast to the metamorphoses in the upper three rows, this process is initially very realistic: the apple is peeled and sliced. These slices join together to form a changed shape and something new emerges. This fantastic apple man, reminiscent of Miro, is capable of various changes in color, shape, and orientation, as the lower row of the illustration shows.

Such metamorphoses as the development of the apple with its drastic changes in shape can also be found in nature: in a similar way a tadpole turns into a frog, a caterpillar into a butterfly, or a seed into a flower.

Now take a look at the last picture in the Fantastic Gallery of Arts. It is a picture by the Japanese artist T. Kuniyoshi from 1990. Figure 4.32 shows a face composed of a whole collection of bodies.

Fig. 4.32 T. Kuniyoshi: a face made of bodies (1990)

Still under the impression of what they have experienced, but also tired, the group leaves the gallery and thus ends their journey through the world of ambiguities.

Arrived outside, the participants look to the evening sky. Slowly, a particularly large cloud moves over them.

"Look, it looks like a Schnitzel with French fries," says one.

"Nonsense. I see a glass of red wine," says another.

Everyone laughs and goes to an absolutely unambiguous pub to end the day with Schnitzel and red wine.

Chapter 5
Fifth Journey: The Colors and the Grey Everyday Life

So close to heaven! Colors, nothing but colors as far as the eye can see. A colorful dream in red, yellow, purple, blue, and green. Balm for the soul, sun for the mind. Sublimity, tranquility, and harmony find their deepest expression here.
By absorbing the fantastic variety of bright colors on a sun-drenched flower meadow, we automatically learn to look deeper, to look behind the cover. As soon as we begin to see through surfaces, we dive into new dimensions: into the world of creativity, fantasy, colors within ourselves.

Let us take you into the enchanting world of our color perception. Marvel at the variety and power of your color vision!

5.1 At Night All Cats Are Grey

"All cats are gray at night!" Surely you know that saying. We will see below that (and why) this is indeed the case. Imagine if it was 6 o'clock in the morning. Outside it is still dark and the moon is shining—a rather unusual starting point for our journey through the world of color perception, one of the most fantastic inventions of seeing! Because in this weak light no colors are recognizable. For example, the attempt to distinguish the colors of the two chairs in the waiting room in Fig. 5.1 fails! If 6 o'clock is too early for you, you can confidently follow the forthcoming considerations at another time. The only condition for this is a room that can be darkened.

T. Ditzinger, *Illusions of Seeing*, https://doi.org/10.1007/978-3-030-63635-7_5

Fig. 5.1 In this picture, the differences between color and black-and-white vision become clear. Look at this picture for some time at dim light. After the two colors appear to be equally bright at first, the blue chair finally looks a lot brighter!

5.1.1 The Purkinje Effect

Take this book with you into this dimmable room—for example into your bathroom, cellar, or storage room, and look at the two chairs in Fig. 5.1 in twilight with the door open. At first, the colors appear intense and similarly bright.

Now close the door very slowly and keep looking at the picture. Little by little the intensity of the colors decreases! As soon as you have closed the door to a narrow gap, you can only see the two chairs in a dim light. Now give your eyes some time to adapt to these changing lighting conditions. In fact, after 2–3 min you will be able to see the two images well—but only in black and white. However, you will notice two significant changes compared to color vision in brighter lighting.

The first difference is that the original blue chair looks much brighter than the original red chair! So blue tones are perceived much more intensively than red tones in reduced light. This effect, the *Purkinje-effect*, was discovered in 1825 and named after its discoverer.

The second difference compared to seeing in bright light becomes obvious when you now try to read the two chair inscriptions or this text. The writing seems very blurred and reading is quite difficult! For comparison, open the door again and continue to look at the writing. Very quickly the visual acuity improves again. But stay in your darkened room until the dog with the purple tail appears!

The result of this experiment is therefore as follows: To be able to recognize characters in a dark place, an enormous visual effort is required, which is many times higher than in normal lighting conditions.

This insight illuminates, among other things, the old argument between parents and children about reading under the blanket and the right time to turn off the light. From the findings of our bathroom experiment, a parental “Don’t read so much anymore, you’ll ruin your eyes” can at least be answered by the children with “Better buy me a brighter bedside lamp or an e-reader!”

5.1.2 Day- and Night-Vision

We can draw some conclusions from our first experiment: Black-and-white vision and color vision are two completely different processes of perception that take action under very different lighting conditions. They also have very different characteristics in terms of visual acuity and brightness perception: Color vision provides a significantly sharper visual impression than black-and-white vision. It requires a lot of light intensity and is therefore mainly used during the day. Black-and-white vision, on the other hand, does not provide by far such sharp visual impressions as color vision. It is less dependent on high luminous intensities and therefore mainly takes over the night shift.

So the old saying that all cats are grey at night is absolutely true. During the day, black-and-white vision is also part of our perception, but it is then completely covered by the color perception.

5.1.3 The Colorful Dog

Our travel group that got up early is still waiting for the bus driver, as a red-brown dog with a purple tail approaches.

"Well, now the night is going to be colorful after all." One of the waiting people is happy.

Thereupon the dog shakes itself vigorously and its violet tail wobbles back and forth like crazy.

You can easily recreate this "shaking scene" in your darkened room. Take Fig. 5.2 with the dog at hand and slowly swing it back and forth in front of your eyes. In fact, the dog's tail begins to wag vigorously, the more you move the picture! If the effect does not occur clearly, reduce the brightness a little by closing the door a little more.

As we will learn later, the reason for this fantastic agility of the picture lies in the alignment of the color of the tail with the background.

Here the variety of colors and their relation with completely different unexpected areas of perception, for example illusory movements, is already evident.

5.1.4 Disappearing Stars

In the meantime, our travel group has started to spend the waiting time for the bus driver by looking at the starry sky. A further interesting brightness-dependent effect can be observed here. It turns out that light-weak stars disappear from our perception as soon as we try to fix them. And they reappear immediately, as soon as we look at something beside the low-light area.

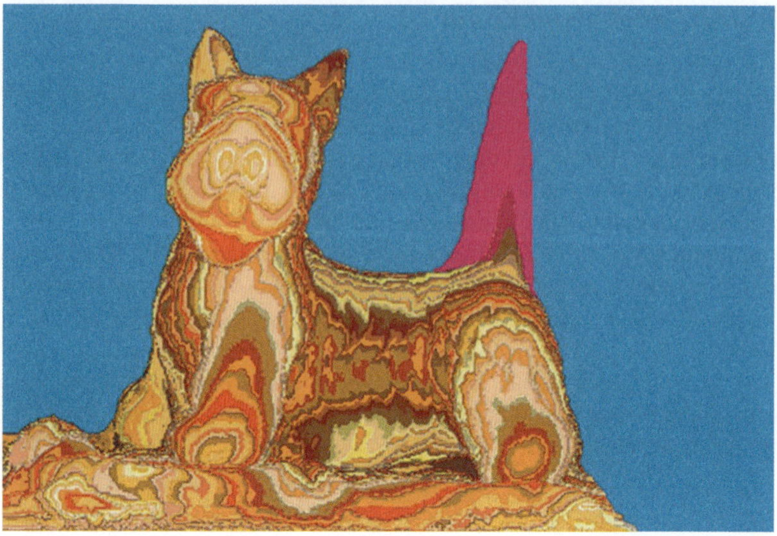

Fig. 5.2 A colored "shaking picture" for weak illumination. Attention: The colorful dog waggles its tail when you shake the book!

This effect, shown in Fig. 5.3, can be tried out very well on a warm summer evening with a view to the stars. Or rather to the light of the stars, for only this we can perceive. To see the real stars, our eyes are hopelessly overwhelmed, because they are far too small. Even if all known star surfaces outside our solar system were joined together, there would only be one disk with a very small diameter of 0.2 arcsec. That's 500 times less than what can be seen with the naked eye! So the only thing we see from the stars is their radiated light.

5.1.5 The Brightness of Stars

Already the Greek philosopher Hipparchus dealt with the different luminosities of the stars. He divided the stars visible to the naked eye into six different brightness classes. The stars of the first class are brightest, and those of the sixth class are just visible with the naked eye. This—the sixth class—corresponds to the brightness of a candle at a distance of 20 km!

The scale of Hipparchus was considerably extended to 24° of brightness after the invention of the telescope.

Stars with a brightness level of 24 can just be detected by photopgraphing them with very light sensitive instruments at long exposure times. Their brightness corresponds approximately to a candle in the unbelievable distance of approximately 100,000 km!

The scale of relative brightness has also been extended to the other side. For example, Vega is a star of zero size. Even brighter stars have negative values. Sirius,

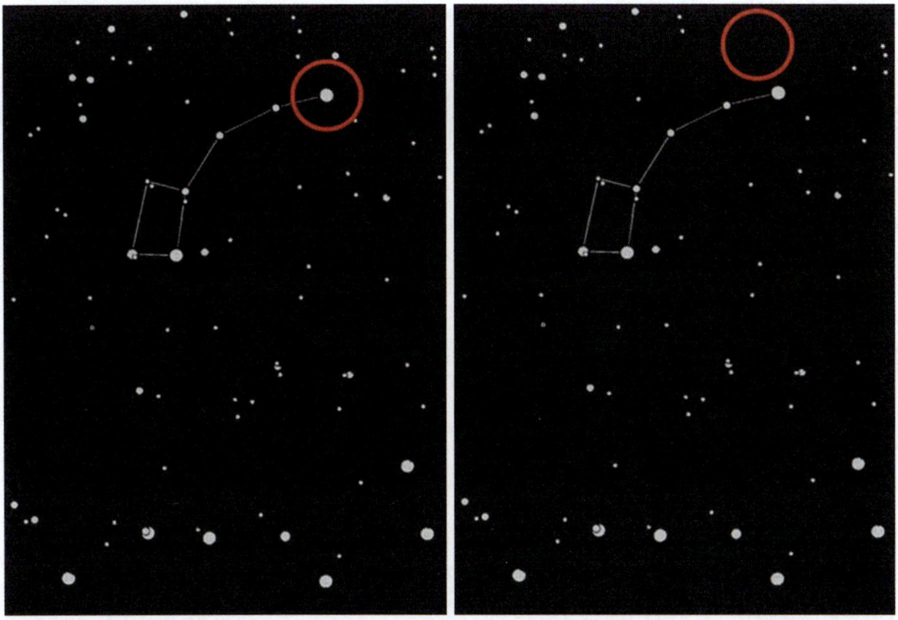

Fig. 5.3 An eye experiment with the stars: Weak stars disappear when you try to fix them—the circle represents the direction of fixation

as the brightest visible star, has a relative brightness of −1.5. For comparison: The sun has a relative brightness of −26.7.

In the nineteenth century, the difference in brightness between the size classes was defined even more precisely: A star of one size class is then exactly 2.512 brighter than a star of the next size class—the reference star is the Polar Star, with the relative brightness of 2.12.

It soon turned out, however, that the Polar Star was not a reliable reference point: its brightness fluctuated slightly. Therefore the following solution was implemented in 1922: the *North Polar Sequence*. This sequence consists of several stars in the vicinity of the northern celestial pole down to the seventeenth size class, which since then has served as a reference system for the brightness calibration of fixed stars.

Today we know that fixed stars such as our sun generate vast amounts of energy, which they release into their environment in the form of electromagnetic radiation in all directions. This radiation spreads across space at a very high constant speed, the speed of light. At the same time, it partly covers immeasurably long distances in unchanged form. Only when it encounters an obstacle such as the human eye, for example, does its long "wandering" come to an end. Radiation from very distant galaxies, for example, has had a travel time of several million years. Nevertheless, we perceive exactly the same light that was sent on its journey from this galaxy before this long time.

5.1.6 Electromagnetic Radiation

The nature of electromagnetic radiation has been understood at the latest with the equations of the physicist James Clerk Maxwell (1831–1879), who described them in an impressive compact way. In order to grasp the full elegance of this description, one must actually read it in its original language, mathematics. Nevertheless, for the sake of simplicity, we would like to try a "translation" of the findings of classical electrodynamics into colloquial language. The following dialogue between the ingenious physicist and his grandmother illustrates the phenomenon of how electromagnetic radiation propagates:

Physicist: "Grandmother, you can imagine it like a telephone: You speak into the receiver, the sound is transmitted through the telephone cable until it reaches your daughter in Miami at the other end".

Grandmother: "That sounds plausible. I'd rather imagine my poodle than my phone: When I pinch his tail in the back, he barks in front. The touch is reported via the nerve conduction to his brain and finally to his vocal chords. But what I can no longer imagine is how the whole thing works in radio?"

Physicist: "It works basically the same way—just without the poodle in between!"

If you do not have a poodle, you will now receive another, more scientific description of the processes involved in the transport of energy by electromagnetic radiation: initially, the energy is emitted in the form of coupled electric and magnetic fields. The coupling is such that the two fields alternate in their strength in waves. It is crucial that these electric and magnetic waves always oscillate perpendicularly (transversally) to their direction of propagation! These *transversal waves* are a very special form of locomotion. The simplest model for this is the wave propagation in a rope, which is shown in Fig. 5.4.

For clarification, two members of the tour group are available to hold a rope as shown in Fig. 5.4. On the left side, the rope is vibrated. Already after a short time the waves arrive at the other end. The oscillations are perpendicular (transversal) to their direction of propagation and arrive unchanged at the right side.

If the rope is only deflected upward and downward, this is a special case of transverse wave propagation: the oscillation is now *polarized*!

If, on the other hand, the rope is also deflected to the left and right, circular wave movements occur around the rope and the transverse oscillation takes place in two planes.

In addition to the polarization of the wave, two other terms play an important role: wavelength and amplitude. The faster the rope is moved, the more wave trains are generated. It is said these waves have a shorter *wavelength*. Of course the rope can also be deflected further. This results in waves with a greater height. This height is called the *amplitude*.

Like the rope waves, the electromagnetic waves also differ in their wavelength and amplitude.

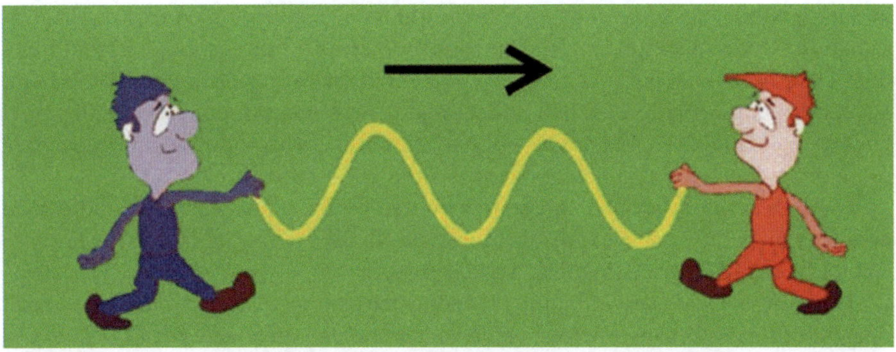

Fig. 5.4 A simple model for a transverse oscillation: Energy is transmitted from left to right. Thereby the rope vibrates upward and downward, perpendicular to its direction of propagation from left to right

Wave trains with different amplitudes differ in the intensity of their visual impression. The following rule applies: The greater the amplitude, the more intense the light appears!

The wavelength of an electromagnetic wave, on the other hand, is directly related to the energy of the radiation. The following applies here: the smaller the energy of the wave, the greater the wavelength, and vice versa.

You have already gained an overview of the entire electromagnetic energy spectrum by looking at Fig. 1.1. It also became clear that we are familiar with the names of most types of radiation in the spectrum. However, the energies of the different types of radiation are mostly unknown to us, since their perception, with the exception of the small range between 400 and 800 nm wavelength, eludes our senses. The visible light is located in this narrow window of the electromagnetic spectrum.

But why can we only see this small part of the entire energy spectrum?

The answer gives us once again the opportunity to admire nature's fantastic quest for an optimum. It is true that radiation from the sun and from the rest of the universe constantly comes to earth from the entire electromagnetic spectrum. But the largest part of this radiation is already absorbed by the atmosphere. Strictly speaking, this natural protective shield only allows radiation to pass through to the earth's surface in two very narrow wavelength windows: One of these is the "optical" window with a wavelength between 400 and 800 nm, the other is the "radio window" with a wavelength between 1 mm and 18 m.

5.1.7 The Visible Light

Our human eye has adapted very well to the natural conditions and "covers" the optical window exactly. Around the optical window, the atmosphere also lets pass some other wavelengths such as ultraviolet radiation (with wavelengths smaller than

400 nm), which can be seen on your own skin while sunbathing. However, this UV radiation is only incompletely transmitted by the earth's atmosphere. Therefore, a consistent light perception in a larger wavelength range (e.g., from 200 to 800 nm) is very difficult to achieve. When evaluating the wavelengths, our perception would have to make too many exceptions—the transmission behavior of the atmosphere in these outer areas of the optical window is simply too different!

The radiation in the second area transmitted by the atmosphere, the radio window, has such large wavelengths that distances in the vicinity of interest to humans cannot be sufficiently distinguished and resolved.

Thus, our eyes are left with only the range between 400 and 800 nm as the visual range. Fortunately, many interesting, characteristic processes take place precisely in this area: Depending on the object, some wavelengths are reflected, some are absorbed and some are simply transmitted. Thus each object appears in its own characteristic light.

This light finally falls on the retina of our eyes. Black-and-white vision, as we know from the first journey, occurs with the help of about 120 million light-sensitive cells, the rods. They consist of a specific visual dye, *rhodopsin*. The coloration of rhodopsin is very clearly visible in flash photography—the eyes are often colored red! Rhodopsin is also called visual purple because of this red coloration.

The rods are by no means evenly distributed over the retina. In the area of the sharpest vision, the fovea, there are no rods—the rods all lie outside this central area of the retina, because this is where the color-sensitive cones are located. This distribution explains why the text in Fig. 5.1 is very blurred in darkened conditions. The phenomenon of the disappearance of the light-weak stars can also be explained from it: If we fix the polar star with our eyes as shown in Fig. 5.3 on the left, the light of some faint stars around it hits the edge of the retina. There are a lot of rods that are able to register even very weak light. Therefore, at the edge of the field of view a (however blurred) perception of such light-weak stars is possible. But if we try to focus our view on one of these stars, as in Fig. 5.3 on the right, the light of this star falls on the center of the retina. Since there are no light-sensitive rods, the weak starlight can no longer be perceived in this case!

5.2 Color Vision

Our travel group now sees things in a brighter light, as both the bus driver and the sun are now there and shining brightly. The bus driver is beaming because he is very well rested after his day off. In addition, the forthcoming fantastic tour through the colors is his personal favorite; after all, there are some wonderful natural beauties on the program today. And the sun is shining—the rain, which was pelting down a short time ago, has just stopped. Immediately a wonderful, colorful, splendid rainbow develops.

5.2.1 The Rainbow

The impressive play of colors of a rainbow has been so familiar to us since childhood that we hardly give much thought to the causes of this natural phenomenon. It would be enough to just be happy about the variety of colors of the rainbow as you can see in Fig. 5.5. But behind the rainbow the fantastic land of colors begins. Therefore, we try to look behind the secret of the rainbow.

The way there leads us through an English market in 1666, where the physicist Isaac Newton bought a special piece of glass: a triangular prism. He was enthusiastic about the way the prism disintegrated the light on one side into all sorts of colors on the other. These colors are exactly identical to the colors of the rainbow. About this phenomenon he wrote a treatise entitled "A New Theory about Light and Colours." In it, he put forward the thesis that white sunlight is composed of all the colors of his prism.

Newton was firmly convinced that this was "probably the most wonderful discovery ever made." But not even with such colorful words could he convince his numerous opponents of the colorfulness of sunlight. As late as 1790, Goethe contradicted him, saying that "the idea of the colorfulness of sunlight is absurd" and that "at most it was suitable as a children's fairy tale."

Fig. 5.5 The colorfulness of the rainbow

In the meantime, however, nobody doubts Newton's assumption anymore: White sunlight actually consists of the whole rainbow spectrum of different colors!

5.2.1.1 How are the Colors of the Rainbow Created?

Raindrops and sunlight are important prerequisites for a rainbow. What happens when the light falls on the surface of a drop of water? Take a look at Fig. 5.6!

One part of the light beam is reflected again by the surface of the drop. This beam is also called a first class beam; the other part of the light penetrates into the drop. This penetrating jet is deflected within the water droplet from the incoming jet in its direction of propagation—it is also said to be *refracted*. Refraction occurs whenever light passes from one medium (e.g., air) to another (e.g., water or glass, as in the prism).

The refraction of light is different at different wavelengths. Thus the energy-rich blue light is deflected more strongly from its original path, while the red light is deflected more weakly. This fanned out ray can now be reflected any number of times within the water droplet and circles the center of the droplet like in a carousel.

At some point, however, the ray leaves the drop again. Depending on how often the beam hits the drop surface, it is referred to as a second-, third-, or fourth-class beam.

The beams of the third class are of particular importance. These form the *main rainbow*, which is most clearly visible. Rays from the fourth class are able to form a *secondary rainbow*, which is much weaker and less visible.

But there is one more difficulty with our explanation of the rainbow. The sunlight is not so narrow that it always falls on a raindrop in the same place. And besides, the rain consists not only of one drop, but of many. Therefore, all possible incident light

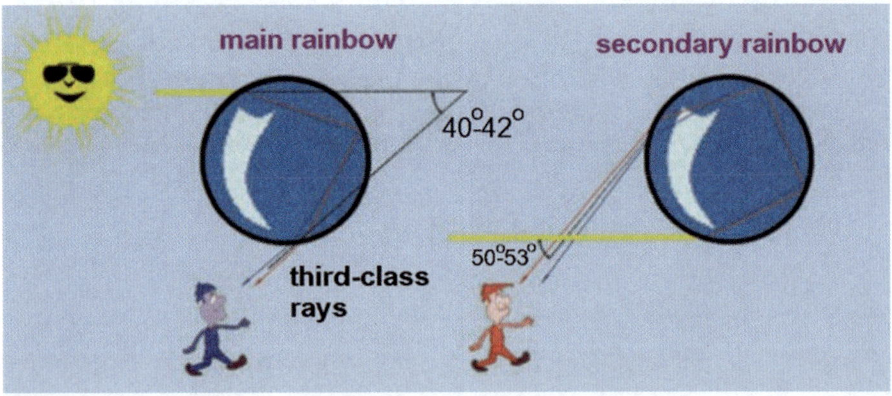

Fig. 5.6 The carousel of light in a raindrop: The sun shines on the raindrop from the left. A part of the jet is immediately reflected again at the surface, while the rest penetrates into the water droplet. When entering the water, the direction of propagation of the jet changes. The magnitude of the angle of refraction depends on the wavelength of the light: blue light is deflected more, red light less from the original direction. This refraction in the drop of water is the cause of the rainbow

beams must be considered separately: For example, if a ray of light passes directly through the center of the drop, the second-class ray is reflected directly back onto the sun. The farther the incident ray is from the center of the drop, the farther it is directed away from the sun on its way back.

The French mathematician René Descartes (1596–1650) was able to show by simple geometric considerations that the majority of all possible rays between 40° and 42° are deflected "away from the sun."

The blue light appears under 40° and the red light under 42°; the angles for the other colors are in between. The philosopher Roger Bacon observed exactly these angles between sun, rainbow, and observer in nature as early as 1266.

The fourth-class rays, which form the secondary rainbow, can also be examined geometrically in detail. Again, it turns out that practically all fourth-class rays leave the water drops at the same angle. These angles are 53° for blue light and 50° for red light. This means that the colors of the secondary rainbow are exactly upside down!

From this, even another insight can be inferred: In the range between 42° and 50° neither third nor fourth class beams are possible! The discovery of this dark area between the two arches is attributed to the Greek philosopher Alexander of Aphrodisias. Therefore the dark area is called *Alexander's dark band*. With the help of Fig. 5.7, you can clearly understand what you have just read.

The spectrum of the rainbow appears so intense because it consists exclusively of pure colors like no other structure in nature!

This means that the light of each color tone consists of exactly one wavelength range—such pure colors are called *monochromatic*. We can make a final statement:

Fig. 5.7 The result of geometric observations of an illuminated drop of water from Descartes: He found the explanation for the main rainbow, which appears between 40° and 42°. He was also able to explain the subsequent Alexander's dark band between 42° and 50° and the sometimes visible secondary rainbow in the range between 50° and 53°, which is colored "upside down"

There is no mixing of light rays with different wavelengths in the rainbow! Our perceptual apparatus is therefore capable of perceiving the monochromatic colors of the rainbow correspondingly different. To this it has a fantastic property that is both useful and often simply beautiful: color vision.

5.2.2 A Connection Between Logic and Emotion

There is hardly a sensory perception where there are so many overlaps and conflicts between art and science, between psychology and physics, as there are with the sensation of color. For example, the effect of colors can sometimes be described much better with "my favorite color" or with "warm" or "cold" than with physical data such as "400 nm." And it is well known that artists are wary of painting according to scientific findings. Vincent van Gogh, for example, said: "The true painters are those who do not paint things the way they are, but the way they feel them."

This dichotomy between feeling and reason is not as clearly expressed in any other sensory perception as in color vision.

The special feature of color perception is that even very small differences in the wavelength can trigger the greatest differences in color perception. These differences can be very different from person to person.

How could color vision develop in evolution at all, and what is its actual practical use?

Through color vision, camouflage is more easily dissolved, structures are better recognized, and prey animals are more quickly perceived, which of course is a very decisive advantage in evolution. Colorful fish in the water or bright red berries in a green tree can be easily recognized. This becomes clear, for example, when looking at the rowan shown in Fig. 5.8. While on the left side in black and white a structure becomes recognizable only after some time, the berries are immediately visible in the identical picture on the right with color.

However, this advantage has to be "bought" with a considerable amount of brain capacity and wiring of neurons. For example, humans have about 6 million colored photoreceptors in each eye—not to mention the cells that process the color information. Due to this high additional effort, nature cracks a walnut with a sledgehammer in a certain way. This also explains why the ability to see colors in evolution has only partially prevailed in some animals or—like in dogs—not at all.

But nature would not be nature if it did not use the enormous potential of its "invention" in other ways. The remaining colored sledgehammer-time, which is not needed for scouting for walnuts and so on, is at our free disposal. This "freedom" results in a slightly different perception of color in every human being. Color vision thus becomes a very personal sensation with different favorite colors and symbols. As with the sense of smell, a direct connection to the subconscious and emotional life is noticeable. Like smells, some combinations of colors appear as pleasant and beautiful or even repulsive.

Fig. 5.8 The Benefit of Color Vision: a Rowan in Black and White and Color

Because of this fascinating connection of color perception on the one hand with the mind and on the other hand with feeling, it is only understandable that both artists and scientists have always felt "responsible" for colors.

However, an adequate description of color vision is probably only possible through a combination of these two areas. This is exactly what we want to try by letting the magnificent colors of the rainbow affect us emotionally once again. At the same time, however, we ask ourselves the logical question of how our perceptual apparatus can manage to perceive so many different colors at all.

5.2.3 *The Three-Color Theory of Seeing*

We humans are able to distinguish about 150 different hues. These hues can also be classified by us as differently intensive and differently bright. This results in the almost unimaginable value of about 7 million solvable different color levels!

How does our eye manage to perceive such an immense amount of colors in high resolution?

This question was also asked long ago by the English physician Thomas Young (1773–1829).

It was clear to him that only a limited amount of space was available on the human retina (approx. the measurement of a postal stamp). More than 100 million photoreceptors with a diameter of 2 μm are needed to achieve our visual acuity! Ideally, around 7 million different color vision cells would be required to recognize all perceptible color gradations at each retinal location. On the retina, however, there is not even room for the 150 different cells to recognize the 150 different hues.

Young knew that nature always strives for the optimum and therefore uses as many different types of color vision cells as necessary—and none more at all. He found out that any shade can be produced by a reasonable mix of a maximum of three basic shades. At first it does not matter which color these three basic tones

Fig. 5.9 The additive color mixing: Colored light, for example from different projectors, mixes and its intensities add up: red and green = yellow, Green and Blue = Cyan, Blue and red = magenta, Red and green and blue = white

have, they should only differ sufficiently. For example, the basic colors blue, red, and green can be mixed to create any other color.

This insight can be illustrated with the help of three slide projectors that throw the colors blue, red, and green onto a screen (Fig. 5.9). The luminosity and brightness of the projectors can be continuously adjusted, so you can mix every possible shade together. If different light rays meet, the result is brighter than the individual components; the intensities of the rays add up. This is why we speak of *additive* mixing. So we get white as a result from red and green and blue. This proves that white light consists of colored light!

All other possible color tones can also be created by changing the light intensity of the projectors.

Three basic colors are therefore perfectly sufficient to represent all the colors visible to us. From this knowledge, Thomas Young drew the right conclusions, which were later extended by Hermann von Helmholtz (1852). As early as 1802, he claimed that there were three light-sensitive structures at every point of the retina that reacted to red, green, and blue. This theory of the trinity of human vision is also known as *Trichromasia* or the Young-Helmholtz theory of color vision (Helmholtz 1911).

The additive color mixture of light is not necessarily familiar to our everyday understanding. The reason for this is that we very rarely have the opportunity to handle colored light rays as we wish. However, it is much easier to mix watercolors or other colored materials with each other and look at their mixed colors. This results in something completely different than the additive mixture of light rays would produce: if you mix yellow watercolor with blue it results in green and not in white! This is the so-called *subtractive* color mixture. The difference to the additive mixture lies in the fact that solid substances (such as watercolor dyes) are mixed together in contrast to light rays.

Although white and green are very different, the subtractive color mixture can still be explained by the additive color mixture.

We want to proceed from the two colors blue and yellow. As you have already seen in the first journey, the colors of an object are determined by the strength of the wavelengths of light it reflects. All other non-reflected wavelengths are absorbed by the object. A blue dye mainly reflects blue and a little in the green range, while it absorbs yellow and red. Yellow watercolors naturally reflect yellow and a little green and red, while blue is absorbed.

But what happens if these two substances are mixed?

The dye mixture removes all colors from the incident light in the area of which it is possible to absorb them. Thus yellow and red are already extracted from the light by the blue dye. In addition, there is blue from the yellow dye.

What's left? Only the green, which is only partially absorbed by both dyes. The remaining green part is reflected by the dye mixture, which results in the green hue according to the subtractive color mixture. The subtractive color mixture is thus created from the additive mixture of the colors reflected by the dyes. What remains is the new color of the subtractive color mixture.

The final confirmation of the three-color theory was not achieved until 1964: Independently of each other, the scientists Paul Brown and George Wald (1964) as well as William Marks et al. (1964) used microscopic methods to study the laws of wavelength absorption of the various photoreceptors. They found out that there are exactly three different color vision cells—the *cones*—which are most sensitive in the vicinity of the colors blue, green, and red. All three cones use the visual dye Iodopsin.

5.2.4 Rods and Cones

In total, there are about 6 million cones on the retina, almost all of which are located in the area of sharpest vision, the fovea. This makes it clear that the color vision cells are responsible for the sharp vision. However, they require much more light than the rods located in the outer retinal regions.

The cones differ from the rods in some essential points, as seen in Table 5.1.

Rod vision and cone vision are possible independently of each other. For example, under poor lighting conditions, only black-and-white vision through the rods is possible. In strong lighting, on the other hand, rod vision clearly fades into the background. The entire visual performance is now provided by the cones. That's why we see much sharper in daylight.

Black-and-white vision is a much more ancient invention of nature than color vision. This is why rod vision is technically not quite up to date. This can be seen, for example, in the resolution of fast image sequences. While color vision can still resolve about 20 frames/s, black-and-white vision is much slower. That is why movements appear faster in low light than in daylight!

The bottom line is that not only all cats are grey and blurred at night. Their movements also appear faster than during the day! Also if one compares the preset image frequency above which no flicker impression is visible, the difference becomes

Table 5.1 Differences between rods and cones

	Cones	Rods
Vision	Color	Black and white
Quantity	6 million	120 million
Dye	Iodopsin	Rhodopsin
Location	In the fovea	At the edge of the retina
Visual acuity	Sharp	Blurred
Sensitivity to light	Low light-sensitivity	Light-sensitive
Time resolution	18–24 frames/s	Slower, movements appear faster than during the day
No more flickering	Maximum 80/s	22–25/s
Processing time of the visual stimulation	Fast	Slow

clear: rod vision no longer perceives flicker at 22–25 light stimuli per second. In contrast, up to 80 image changes are required for cone vision.

The example of the bee shows that also much higher time resolutions of image sequences are possible. The bee can still resolve 100–200 frames/s.

This value is even too high for our TV and computer screens with 50–100 frames/s. That's why our television appears to the bee only like a slide show of standing, changing images.

Another proof that rod vision is not up to date is the processing time of the visual stimulus.

Black-and-white perception requires much more time in its processing than color vision. This time difference can be illustrated by the *Pulfrich effect* (Pulfrich 1922). This effect is a deception of perception when looking at a swinging pendulum. Let a pendulum swing in front of your field of vision. If you now weaken the brightness in front of one of your eyes, for example by holding one screen of your sunglasses in front of one eye, the information about the position of the pendulum will arrive somewhat delayed compared to the other eye. That is why the pendulum motion no longer appears in one plane, but elliptical.

5.2.5 How Does Color Vision Work?

The photoreceptors act as "catchers" of light. They prefer to collect the light that suits them best. If the wavelength of the incident light is exactly within the range of the maximum sensitivity, the light is fully registered. However, the more the wavelength deviates from this point of greatest sensitivity, the weaker the respective photoreceptor evaluates the incident beam. If the light wavelengths deviate too far, the sensitivity of the visual cells is zero, i.e., they no longer react to the incident beam.

The three different types of cones have clearly different sensitivity ranges, which overlap and cover the entire spectrum. The sensitivity of the "blue" cones (left curve

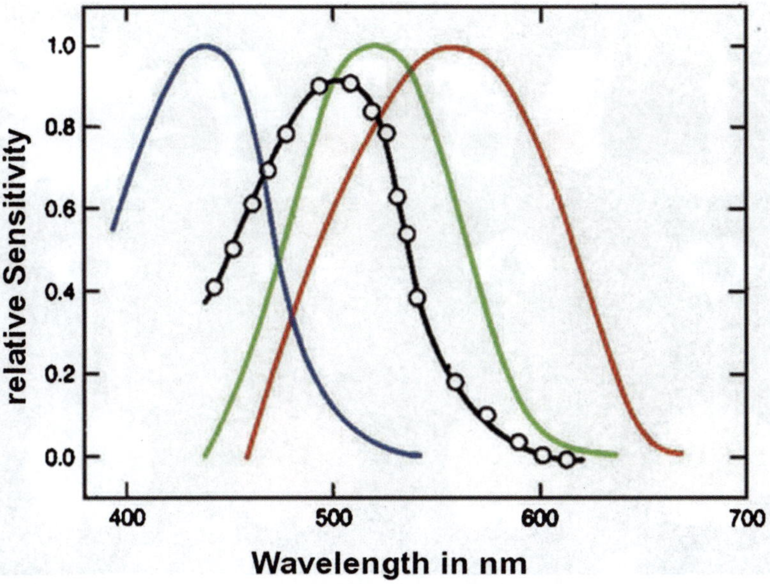

Fig. 5.10 The sensitivity of the cones (blue, green, and red curve) and the rods (black curve) as a function of the incident light wavelength

in Fig. 5.10) starts at approx. 400 nm and increases with the wavelength. With the color violet blue (448 nm), the sensitivity reaches its maximum. Then it sinks again until it disappears completely at approximately 550 nm.

The same applies to the curves for the two other cone types as you can see in Fig. 5.10. The mean sensitivity curve of the "green" cone has its maximum at about 518 nm, and the "red" cone at about 617 nm, which corresponds to the color orange-red.

The different photoreceptors therefore simply measure how well an incident light beam matches their own color. If they register a good match, they produce a very high stimulus, otherwise a low one. This produces three nerve stimuli of different strengths, which are evaluated in the visual center to form a single color impression.

For example, an incident beam with a wavelength of 580 nm, as shown in the left of Fig. 5.11 is not registered by the blue cones at all. On the other hand, the green and red cones strike approximately equally strong. These stimuli are then processed in the visual center to produce the yellow color impression.

Apart from the monochromatic colors of the rainbow, natural light very rarely consists of only one wavelength. Rather, there are usually many different wavelengths involved in a color impression.

The same color impression of yellow occurs, for example, when we mix red and green light equally strongly (second case in Fig. 5.11). This time the red cones are stimulated by the red light and the green cones by the green light. From this, our

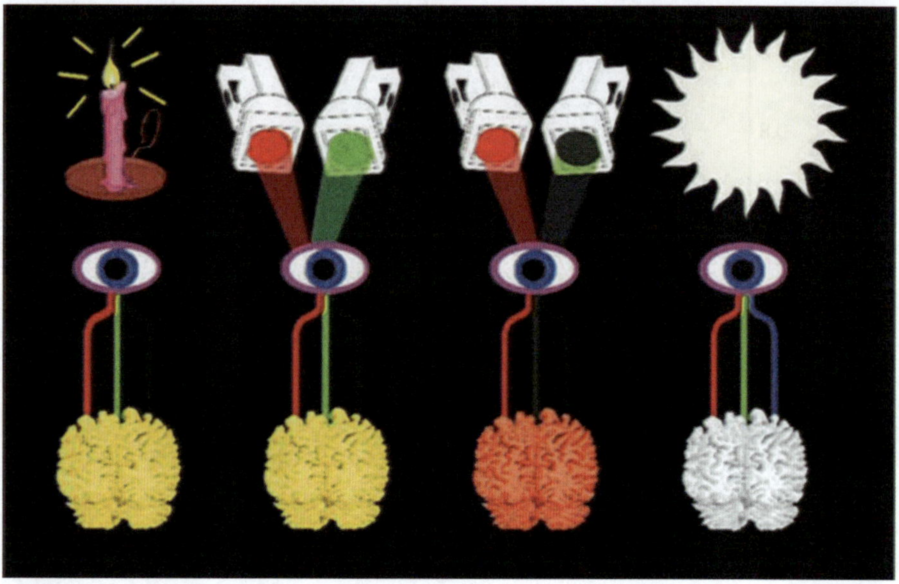

Fig. 5.11 Four tasks for color vision

visual center calculates the color impression of yellow—even though the incident light beam has a completely different composition.

Of course, the cones also react to different intensities. If, for example, a very bright light beam falls on the retina, a much stronger stimulus is transmitted to the visual center. What happens, for example, if the incident light still consists of the colors red and green, but no longer in equal parts, i.e., in a ratio of 10:1? This can be seen as the third case in Fig. 5.11: This stimulates the red cones considerably more than the green ones, which is evaluated in the visual center for the orange-red color impression.

The evaluation of the sunlight is also done in this simple way via the three different stimulus receptors. The solar spectrum extends over all wavelength ranges of the visible range in approximately the same strength. This means that all three cone types are excited in equal parts. These equally strong stimulus receptions lead to white color stimulation in the visual center, as can be seen in the latter case in Fig. 5.11.

What is the relationship between black-and-white vision and color vision?

First of all, it is important to know that black-and-white vision is a physically independent visual impression that has no direct influence on color vision in the visual center. Nevertheless, a comparison can be made between the day and night vision methods: The sensitivity of rod vision—schematically shown as the black curve in Fig. 5.10—reaches its highest value at a wavelength of 505 nm. Incident light of this wavelength is therefore most strongly perceived by the "nocturnal" rods.

We can compare this value with the value of the total sensitivity of color vision. The curve of the overall sensitivity of all three cones together is largely similar to

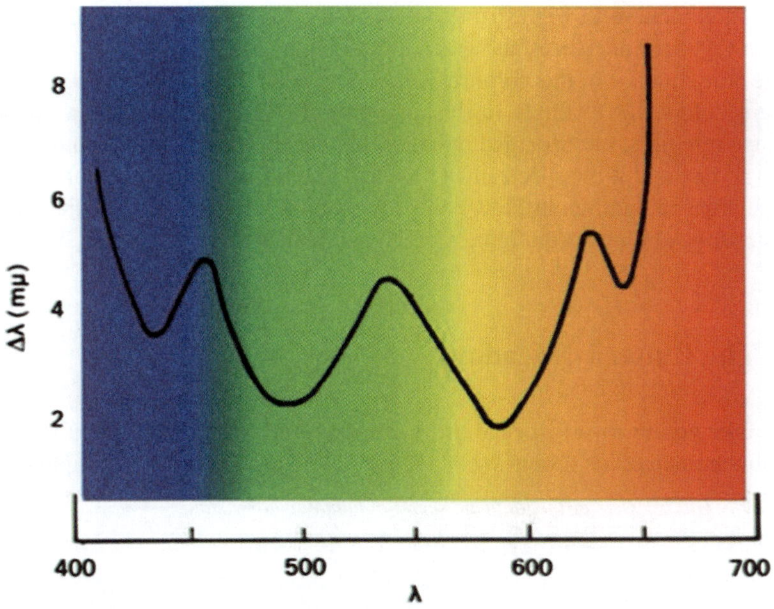

Fig. 5.12 Color resolution curve according to Gregory

the curve of the central green cone vision. It has its maximum at about 555 nm. Seeing in daylight therefore favors the shade of green.

This difference in the wavelength of the maximum sensitivity between brightness- and twilight-vision is the cause of the Purkinje effect, which we have already seen in Fig. 5.1. Green and red are therefore placed brighter in daylight compared to blue than in low light!

Figure 5.12 brings us one step closer to understanding our color vision. It shows the color resolution curve of an average person (after Richard Gregory 1997) and the extremely non-linear resolution of our cones as a function of the respective wavelength. The resolvability is the smallest wavelength difference that still produces a noticeable difference in hue. An explanation for this wild ascent and descent of the curve can be obtained from the comparison with the cone characteristics in Fig. 5.10: The color sensation changes little as long as the wavelength at the ends of the spectrum of Fig. 5.10 is varied. The further we get into the middle of the spectrum, the more complicated the whole thing becomes. If we move slowly from the outside into the increasing absorption areas of red or blue, there is a steady increase in resolution—as long as there is no interaction with the other color absorption areas. In the medium wavelength range, the resolution effects of the individual absorption curves overlap and even a small shift in the wavelength can significantly change the states. The color resolution of the individual colors is best at the points where the corresponding sensitivity curves have their greatest gradients. The best resolution effect is obtained when the relevant gradients involved are opposite, e.g., yellow (green curve in Fig. 5.10 decreases and red increases) or blue-green (blue curve in Fig. 5.10 decreases and green and red increase).

One consequence of this color resolution characteristic is that we do not perceive the pure colors, for example in the rainbow (cf. Fig. 5.5 right), as a continuous sequence of colors, but arranged in bands. While in the low-resolution regions (the bands) at the ends of the spectrum at blue and red or between blue-green and yellow in large areas, a continuous slow color transition is visible; between the bands the colors change abruptly from blue to green and from yellow to orange.

5.3 The Butterfly Meadow

Let us take you to a very special green meadow full of colored butterflies to illustrate how human color vision works. There's a big sign with this inscription:

Butterfly Hunt: Discover the fantastic secret of human color vision!

5.3.1 The Equipment

The bus driver has already prepared three butterfly nets of different colors, which he distributes to three volunteers. The three nets can be seen in Fig. 5.13.

The blue net is very large and has large meshes, the green net is medium size with medium mesh spacing, and the red net is small with correspondingly fine meshes.

From his long experience, the bus driver knows that the upcoming butterfly hunt is one of the highlights of every journey into the kingdom of colors. Besides, the butterflies are still an excitement for him too, as they are always good for a surprise from trip to trip. It is not for nothing that butterflies are the prime example of triggering chaotic processes. It is well known that chaos research claims that a wing beat of a butterfly in Beijing, for example, can have a decisive influence on the weather in our country.

The butterflies on this special meadow are available in all pure colors, i.e., the colors of the rainbow, and in various sizes. They vary in size depending on color: the blue butterflies are the largest, they just fit into the large blue net; they are too large for the two smaller nets. For the smallest butterflies, the red ones, the exact opposite is true: they can only be caught with the small red net. They slip out again through the big meshes of the other two nets. Likewise, the medium-sized green butterflies only get caught in the green net because they slip through the meshes of the large blue net and do not fit into the small red net.

Fig. 5.13 The butterfly nets

5.3.2 *The Butterflies, Einstein, the Light, and the Colors*

These butterflies are a good model for light and its colors. As you already know, light is nothing more than an electromagnetic wave that has been described in detail since Maxwell. In this book we have always understood light as a wave. However, light does not always behave as a wave. Sometimes the "behavior" of light can only be explained by imagining it as a collection of particles—these light particles are called *photons* (Greek: "photein"—shine).

The various ways in which light can be described have already led to very violent disputes in history. For example, Newton, already known to you, represented the particle hypothesis, while Young proclaimed the wave character of light. It turned out that this ambiguity of descriptions was not resolvable for light. A fascinating compromise was therefore reached: Sometimes the light behaves like a particle and sometimes like a wave! Only the knowledge about both possibilities—also called *wave-particle dualism*—completely explains the phenomenon of light.

The easiest way to explain color vision is to consider light as an assembly of particles. These photons of different colors differ in their energy or, as known since Albert Einstein, in their dynamic mass. The energy-rich blue photons have the highest mass and are therefore the largest. The green ones are medium in size and the red ones are very small.

Thus the analogy to the colorful butterflies is obvious. The butterflies are a vivid model for the photons and our butterfly nets illustrate very well the color perception in the human eye.

The bus driver also points this out and claims: "That's why I did all the work to paint the butterflies in different colors according to their size!" At the same time, he

Fig. 5.14 Colorful hustle and bustle on the butterfly meadow

once again dreams of the moment when he is no longer dependent on sayings of this kind and the corresponding tip.

The entire activity on the butterfly meadow is summarized in Fig. 5.14.

5.3.3 *The Color Black*

"Somehow I am pessimistic that something exciting is about to happen here," complains one of the participants after some waiting on the meadow.

The bus driver, on the other hand, countered: "Well, listen! That's a great start, that none of us has a light butterfly in the net. And zero times blue and zero times green and zero times red gives the color black."

5.3.4 *The Color Red*

"I'll soon see red if nothing else happens here," the same traveler whimpers impatiently. And behold: Suddenly a small red butterfly (see Fig. 5.14 above) actually flies up from the meadow and is caught very easily by the red net.

"Red. Right, we see red." the bus driver is pleased.

Afterwards five more red butterflies fly up (see second picture in Fig. 5.14), which are all collected again by the red net.

"Now we see a very strong red, the *intensity* has clearly increased," cheers the bus driver.

5.3.5 *The Color Yellow*

Now the butterflies really come into their own. Next, four butterflies fly up: two green and two red (see Fig. 5.14, third picture). Now, besides the red net, the green net also takes action and quickly collects two medium-sized green butterflies. As you already know from Fig. 5.11, an equal number of the colors green and red results in the perception of the color yellow.

To prove this, four yellow-colored butterflies ascend once more. They are halfway between the red and green ones in size and fit into both the green and the red nets, although they are a little too small for the green net and a little too big for the red net. So both volunteers have some trouble catching the not-quite-fitting butterflies. But they'll make it.

Since they are both about the same speed, they finally each have two yellow butterflies in the net—the same result as before with the two red and two blue butterflies. This makes it clear that yellow can be a pure color on the one hand, but also a mixture of the pure colors red and blue, on the other hand.

5.3.6 The Color Magenta

In the next demonstration, two big blue butterflies fly up, together with two small red ones (see Fig. 5.14 below). Very quickly they land in the corresponding blue and red nets. Blue plus red results in the color magenta. Therefore a swarm of magenta-colored butterflies should fly up now for confirmation.

But nothing happens. How is that possible?

The simple reason for this is that there is no pure magenta or purple in nature! In the rainbow the scale ends on one side with blue and on the other side with red. The color magenta is therefore merely an "invention" of our tricolor vision system. The spectrum of the rainbow is bent by our visual center to a color circle as shown in Fig. 5.15. The connection point between red (A) and blue (B) is continuously filled with purple and magenta tones.

This realization that our perceptual apparatus can recognize these entirely new, unnatural hues such as magenta and purple is a further proof of the three-color theory of seeing. With the help of the butterflies this becomes easily understandable. If these newly "invented" colors did not exist, many medium-sized butterflies would have had to fly up instead of the two large and two small ones—and these would have the color green.

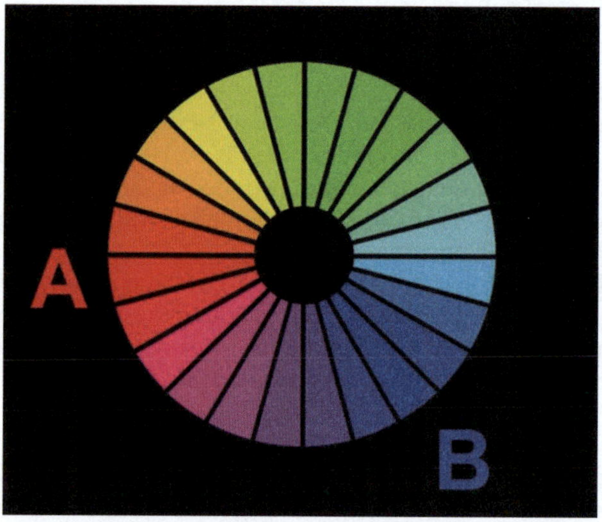

Fig. 5.15 Our perceptual apparatus orders and supplements its colors: The pure colors of the rainbow end on one side with red (A) and on the other side with blue (B). The color gap fills our color perception with new, freely invented colors. These "artificial" colors in the range between A and B are mixed colors between red and blue

5.3.7 *The Color White*

Meanwhile all butterflies fly up in different colors at the same time and the three catchers have their hands full to catch them again with their nets. Since everyone was equally fast, there are about the same number of butterflies in each net at the end—this results in the color white in the visual system.

Just when all the butterflies have been caught, five more red butterflies fly up from the neighboring meadow, but they are quickly caught in the red net. This means that there are five butterflies more in the red net than in the others, but this is not particularly noticeable. In contrast to the demonstration of the color red (see above), where the five individual butterflies led to an intensive red, the result is now a weak, very bright red.

In addition to the intensity of a color, the brightness—that is, the added white portion—is also decisive for the color impression.

5.3.8 *The Complementary Color to Red*

The butterflies have now all escaped and flown. The bearers of the blue and green net are busy trying to catch their butterflies again. At the same time, the bearer of the red net is quite exhausted because there have been so many red butterflies to catch so far for him. The result is that the blue and green nets are quickly refilled, while the red one is much emptier. Thus the white light is perceived as cyan green—this is the *complementary color* to red! The word complementary color comes from Latin ("complementum") and means complementary color. The complementary color is the color that adds another color to white.

This concludes the presentation of the butterflies on how color vision works and we continue our fantastic journey through color perception. After this exhausting demonstration, the butterflies now rest a little and prepare for their next demonstration in Beijing for a travel group on an adventure trip through Chaos research.

5.4 A Round Trip Through Color Vision

The last demonstration of butterflies leads us to a phenomenon that we have not yet paid any attention to: *color adaptation*.

5.4.1 Color Adaptation

Illuminate the red color field in Fig. 5.16 with bright light. Now look at the red color for about 45 s by holding your right eye very close to the book. Close your left eye at the same time.

After this time, the red cones of your right eye are just as overloaded as the red net on the butterfly meadow. That's why you can't perceive intense red tones with the right eye for some time! You can check this very easily by alternately looking around with one eye and holding the other. It becomes clear that all reddish or orange objects appear much more saturated and brighter with the fresh left eye! Only after some time does the red visual pigment in the right eye rebuild and the visual performance of the two eyes is approximately the same again.

This adaptability of the eyes is a great advantage for us! This becomes clear when we imagine other lighting situations. For example, let us think of a room that is illuminated by the yellow shimmering light of a lamp. After some time this light automatically appears as white to us. The eyes have become accustomed to the new environment and things appear to have their familiar colors again. Likewise, the eyes adapt to the coloring of yellowish ski-glasses or sunglasses. Very soon we will no longer be able to see any color differences compared to our normal vision. But if one would take a color picture through these sunglasses, a clear color cast would be recognizable.

This phenomenon of color adaptation of our perception is called *color constancy*. It is true that the cones objectively perceive very different visual stimuli compared

Fig. 5.16 Check your color matching ability! To do this, look at the red color with the right eye from a short distance for about 45 s, keeping the left eye closed! Then look alternately with both eyes at your surroundings. You will immediately notice the difference: You will now perceive red objects much more intensively and brightly with your left eye than with your right. If you look at a white wall, you can see the complementary color

to normal sunlight. However, due to the property of color constancy, our visual center is able to adapt perfectly to new lighting situations.

Of course, color constancy has its limits. If, for example, the color composition of the light source is too far away from the solar spectrum, misjudgments will inevitably occur—even after a very long adaptation period.

Perhaps you have wondered about the strangely changed colors of your clothes under the orange light of a street lamp. The reason for this is that the blue tones are completely missing here. This makes the blue cones "unemployed" and the generation of the color sensation is only up to the green and red cones.

5.4.2 Disorders of Color Vision

Similar conditions are present in persons with color vision disorders. Thereby normally one type of cone is either completely or partially nonfunctional. Color vision disorders are mainly genetic and normally occur in both eyes together.

Studies have shown that 7–8% of men completely or partially lack color vision. In contrast, the proportion of color vision defects in women is only about 0.3%. The reason for this is that the main cause of color vision defect is located at the X chromosome and is inherited recessively.

Women are known to have two X chromosomes. For this reason, a woman is only color-faulty if both X chromosomes have a tendency to defective vision—which is quite rare. As soon as one X chromosome is involved without a tendency to color-ametropia, it dominates the other. A woman who has a disorder of ametropia on an X chromosome may then have normal vision, but can pass the disorder on to her children or grandchildren.

Men, on the other hand, have only one X chromosome, which they inherit from their mothers. The balancing second X chromosome is missing! For this reason, any inherited color vision defect from the mother comes into effect completely.

The largest proportion of ametropia involves a defect in the functioning of the green cones (red-green blindness, which affects about 6% of all men), followed by the red cones (about 1.8% of all men). In contrast, a defect of the blue cones, which are located on another gene, is very unlikely (0.005% of all men).

There are various strong defects in the functioning of a cone. In the most harmless case, only the maximum sensitivity of a cone is shifted with respect to the normal value. But you can still see with all three types of cones.

The next worse color vision defect is the total loss of one type of cone—only two cone types remain for color vision. Persons with such a visual impairment—the *dichromats*—are unable to assign all colors to a clear sensory impression. That is why they do not see any difference between some colors—depending on which cone type is not working. This mistake can be very well determined with color plates called *pseudoisochromatic* plates. They show letters or numbers with different colored spots, all of which have the same brightness.

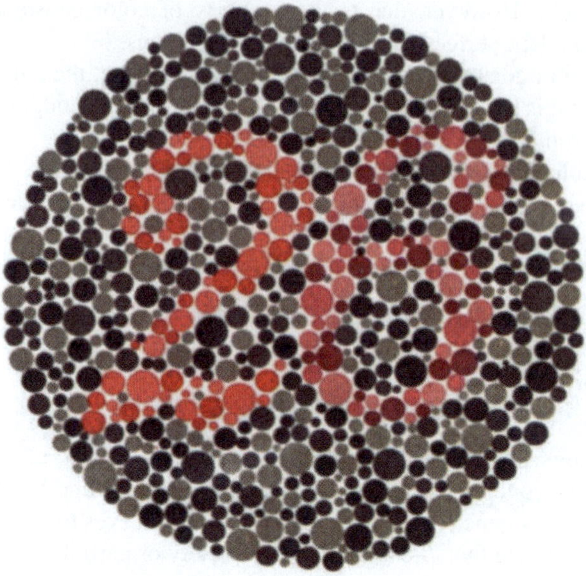

Fig. 5.17 Which number do you see? Example for a color test with the pseudoisochromatic Ishihara plates (Ishihara 1959). *Test resolution*: 26 (normal color vision), 2 (failure of green vision), 6 (failure of red vision)

Check your color vision at Fig. 5.17! If you see a "26," you are capable of normal color vision. But if you see only one number, you are a dichromate: If you see only a "6", your red cone function is missing, if you see a "2," the green one.

Many people do not even know that they have color vision disorder and are accordingly surprised when they learn about it.

The reason for this is that they have learned to deal with this deficiency since birth and they have never consciously noticed it. For example, people with a color vision deficiency unconsciously adapt to the designation of colors by the normal-sighted majority, even though these colors actually evoke a completely different sensory impression in them. Although there are some "ambiguous" colors in their vision, they are still able to assign them to distinct impressions through everyday experience. Only the viewing of such color plates as in Fig. 5.17, which offer no further clues for finding the numbers except the color, does not allow their perceptual apparatus to cheat.

A negligibly small proportion of people (approximately 0.01%) are completely color-blind. In this case, two or even three cone varieties are inoperable.

Seeing with only two different types of cones therefore entails only a very slight disadvantage for the people concerned, which they normally do not even notice.

The only problem is that some colors, which in reality are different, may have identical sensory impressions.

Let us consider the most common failure of the green cones: What is perceived in such a case when violet light hits the retina? Both the red and the blue receptors are receptive. Since these receptors are already all functional photoreceptors, it is no longer possible to differentiate between white light: white and violet in this case produce the same sensory impression!

Such an ambiguity, however, does not hurt much, because our perception is blessed with a variety of reserve recognition systems: The laws of seeing as well as the knowledge of forms and shapes remove this ambiguity with ease. It is therefore no wonder that those affected are often not aware of such visual impairments. How should they know that the violet flower looks different than the white wall behind it, when they have been used to seeing things this way since birth?

In the animal world, too, the ability to perceive color is not always fully developed, as a glance at the following table shows. In vertebrates, the sense of color is probably only present in individual groups.

The example of the honey bee shows how different the color perception of animals can be even with very well-developed color vision (Table 5.2).

5.4.3 The Fantastic Color World of the Honey Bee

Like humans, bees have three different types of cones. However, their highest sensitivities are to be found in completely different areas: in the colors ultraviolet, blue, and green. Bees therefore have no possibility of perceiving red light, but one in the ultraviolet range.

Most of the objects in our environment are active in both the ultraviolet and red areas. Depending on their biological or chemical composition, they absorb or reflect portions of sunlight in their characteristic way. Therefore they can look completely different for bee eyes than for human eyes.

The chlorophyll "green" of the leaves is very important for bees. The chlorophyll absorbs light from the red range. Everything else is reflected. This is why people perceive the complementary color green. But what does the bee eye see? Since the chlorophyll also reflects the UV light, all wavelength ranges visible to the bee eye are reflected. Therefore the green of the leaves appears to the bees more or less as white. The sensitivity of the bee cones is therefore ideally matched to their natural environment. Against the natural background of the "white" meadow, the bee's eye can see things deviating from it most clearly!

The coloring of the flowers is particularly important both for the bees and for the reproduction of the plants. In order to be distinguishable from the white background

Table 5.2 The different color vision of animals

Mice, dogs, rats, rabbits, monkeys, nocturnal animals, amphibians, reptiles	Only weak or no color perception
Cats	Probably dichromats
Squirrels, primates, fish, apes, birds, bees, flies	Highly developed color vision

of the meadow for the bees, they have adjusted to the bee eye in a co-evolution and absorb in at least one wavelength range visible to the bee. A side effect of this coloring in order to attract bees is that these colors are also beautiful for the human eye. For example, bees see a color that we humans see as red, ultraviolet, or black; bees see a color that we see as blue as a mixture of blue and ultraviolet.

But bees do not only succeed in finding the most diverse flowers and plants on the meadow with ease. They can also easily recognize other objects because of their perfect color matching on the meadow background.

For example, while it is difficult to impossible for us humans to discover a lost green key folder in a green meadow, a bee can do this very easily with the help of its UV vision. Since the folder normally does not reflect ultraviolet light, it stands out clearly against the white of the meadow for the bee.

By the way, bees are one dimension ahead of humans when it comes to the world of colors: they are able to perceive the polarization of light. This ability is invaluable for their orientation. Daylight is polarized differently depending on the position of the sun. That is why bees are able to determine the position of the sun exactly and orient themselves to it, regardless of the weather.

This additional sensory impression is unfortunately not possible for us humans. If we could see the polarization of the light, this would be comparable to the change from black-and-white to color vision.

This example from the animal world makes it clear that colors are always relative. There is no fixed system of reference, and every living being has its own individual perception of color, at least in nuances.

5.4.4 The Negative Afterimage

Look at Fig. 5.18 for about 1 min with a fixed gaze. Fix the cross in the middle of the picture. I'm sure you'll find it's not that easy. Instead of fixing the cross, one has the feeling that one's eyes are constantly wandering back and forth in the picture. This is a special protective mechanism for our retina: Called *saccadic eye movements*, the aim of these involuntary jerky eye movements is to protect our retina from the dangers of the very effect that we want to see in Fig. 5.18. Saccadic eye movements prevent a photoreceptor from being overloaded by the same color stimulus. This constant change of the field of view is comparable to screen savers for computers, which are also intended to prevent the screen points from being overloaded or burned in by the same light beam.

This protective mechanism of saccadic eye movements makes it difficult for us to fix the cross for a longer period of time. Nevertheless, the colored areas soon become considerably paler. Only at the transitions between the surfaces do they remain intensely luminous. At these edges the picture even becomes almost alive. The saccadic eye movements play a very important role here: They ensure that the photoreceptor area that perceives the edges is supplied with both colors. Therefore,

Fig. 5.18 Negative afterimage: Fix the cross in the middle for about 1 min. Then look at a white wall. You will now see the image in the complementary colors that are more familiar to you

these photoreceptors are not saturated, but rather there is a high-contrast constant flickering of the edges in the colors involved.

After 1 min of fixation time you look against a white wall. You can immediately observe the negative afterimage. You can clearly see the image in the familiar complementary colors: White instead of black, blue instead of yellow, magenta red instead of green, and so on.

Just like the carrier of the red net on the butterfly meadow, the cones are tired because of the previously burned-in color. Therefore they are not able to achieve the same performance as the other, unused cones with the now incident white light. This creates the impression of white light minus the burnt-in color—and that is the complementary color!

You can tell that this negative afterimage is not a real image by the fact that the image moves with your gaze when you look at another section of the wall!

With this afterimage you can observe a very important characteristic of our perceptual apparatus, which plays a role both in geometrical-optical illusions (cf. the second journey) and in size and depth perception (cf. the sixth journey): *size constancy*. The size of the afterimage varies depending on how far away the white wall is. Although the afterimage in our field of vision is always the same size, this size is evaluated quite differently by our perceptual apparatus by comparison with the environmental conditions. If we look at a close wall, the afterimage appears very small. Conversely, the image appears large on a distant wall.

Such afterimages convey a very special appeal, as they give us the direct opportunity to observe the functioning of our visual system. By the way, this fascinating mixture of science and joy long ago delighted Johann Wolfgang von Goethe (1810):

> *When I entered an inn towards evening and found a well-grown girl with a blinding white face, black hair and a scarlet bodice, I looked at her, who was standing in front of me at some distance, sharply in the twilight. As she now moved away, I saw a black face on the white wall facing me, surrounded by a bright glow, and the rest of the clothing of the perfectly distinct figure appeared to be a clear sea green.*

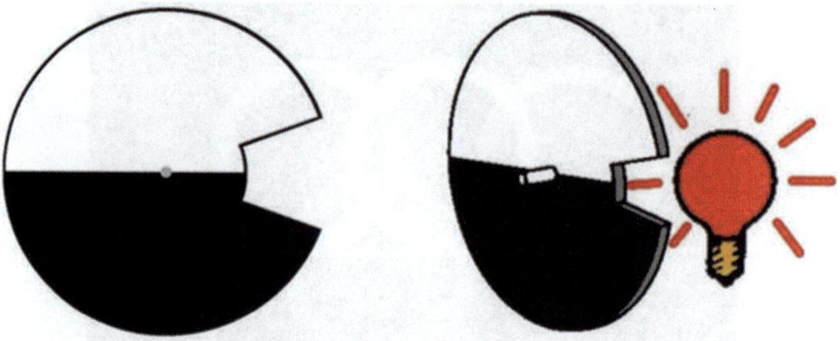

Fig. 5.19 The Bidwell disc: Rotate this disc and look at the objects behind it through the cutout. The trick works best when the objects are illuminated brighter than the pane itself or when they shine themselves, such as colored light bulbs. If you turn the disc clockwise at the correct speed, the disc will appear in the complementary color of the object

This quotation leads to the conclusion that Goethe found a healthy mixture of research and pleasure in his observations. No wonder he wrote an entire book on the theory of color!

A similarly colorful mixture of science and joy created the items we will encounter in Figs. 5.19 and 5.20.

5.4.5 *Rotating Discs*

The whole trick with Figs. 5.19 and 5.20 is that you have to rotate the discs shown. If you do not want to use the book as a Frisbee disc, it is best to copy the page, stick it on a firm base (for example a cardboard box), cut out the circles, pierce the middle with a toothpick, thus creating a rudimentary gyroscope.

Or you can put the circles on an old rotating turntable player if you still have one in your storage room!

The so-called Bidwell disc (Bidwell 1899) with the lateral recess in Fig. 5.19 occupies a certain special position. Cut out the recess and turn the Bidwell disc clockwise into fast rotation (a few revolutions per second!). If the speed is correct, you will see the complementary color of the object over which the disc is circling.

This effect can be explained with the help of the knowledge we have gained about negative afterimages. Through the recess you can see the color of an object behind it—for example, a red tablecloth. The consequence then: The red cones saturate somewhat.

When turning clockwise, the hole is then covered by the white surface of the disc, creating a cyan afterimage. If the rotation speed is correct, the afterimage will light up longer than the red output stimulus and the disc will appear cyan.

If the rotation is reversed, i.e., counterclockwise, there is no possibility to form afterimages, as the black surface initially covers the object under consideration. By

Fig. 5.20 Benham discs: If you rotate these disks, the black-and-white patterns become colored sectors or colored concentric rings by themselves. The colors depend on the direction of rotation and the rotational speed of the disc. The rotational speeds of an old turntable are normally sufficient for the development of the colors

the time the white area is at the same location, half a revolution has passed and the red visual dye has refreshed again. Therefore the Bidwell disc appears red when turned counterclockwise!

While it was always possible to provide a scientific explanation of the observed phenomena in the last few experiments, we will reach the limits of our ability to explain them in the next experiments. The discs in Fig. 5.20 display the so-called *Benham discs* (Benham 1894); these were a popular toy in the nineteenth century. While the Bidwell disc needs a color from the outside and transforms it accordingly, the colors on the Benham disc are created by themselves—only from the black and white original.

Rotate the discs shown in Fig. 5.20 one after the other and observe the resulting play of colors!

These color appearances result exclusively from the temporal change of black and white, which in turn suggests an afterimage effect. But the Benham effect is even more than the usual fixed afterimage. The key to understanding the mysterious disc colors is probably the different stimulus transmission times of the different color sensations in our perceptual apparatus.

The white light, which flashes differently depending on the position of the retina, stimulates all cones equally at first. However, the reaction time and the stimulus transmission time of the different cones are not the same. The original white light is broken down into its components according to the blinking stimulus. By the way, it's noticeable with the different discs that the color pink appears again and again.

The fact that the duration of stimulus transmission is indeed different for different colors can be impressively verified by the following Phenomenon of the Fluttering Hearts.

5.4.6 *The Phenomenon of Fluttering Hearts*

Take a look at the colorful house in Fig. 5.21. What happens to this house during an earthquake can be understood with a simple experiment: To do this, darken your room and move the house in "earthquake waves" from left to right and back in front of your field of vision. If you do everything right, the two windows in the right part of the house seem to fall down immediately and the smoke from the chimney begins to wobble. On the other hand, the left part of the house and the roof are very solid. If you look closely, you can see at least the left door handle wobble as well.

The discovery of this fantastic apparent mobility of different colored surfaces is attributed to the British physicist and inventor Charles Wheatstone. In 1844 he observed the apparent movement of a red-green patterned tapestry illuminated by a flickering gas light. Helmholtz called this effect the "fluttering heart." To explain this phenomenon, we refer again to the different adaptation times and processing times of the different color stimuli.

In good lighting conditions, our color perception is easily able to keep abreast of all shaking speeds of the image. But this changes drastically when we watch the same scene in darker light. The reason for this lies in the significantly slower stimulus transmission time of all the cones involved. The color stimulation processing can no longer keep up with the shaking movement in terms of time. As a result, there are discernible differences in the processing speed of the individual color channels visible. Suddenly some colors start to swing against each other! This can be observed very clearly in the combination red/green and in the smoke in the combination magenta/cyan. On the other hand, the left part of the house with the combination blue/yellow does not swing at all!

Fig. 5.21 A normal house—but only as long as there is no earthquake! If you move the picture quickly back and forth in dim light, parts of the house begin to move. In darker light the right windows, the smoke from the fireplace and the left door handle shake!

It is known that blue is processed much more slowly in the human perceptual system than, for example, green and red. It looks quite like that only those colors "swing against each other" that have a comparably long processing time (Ditzinger et al. 2000). In reality it is not a matter of "swinging against each other" but of "swinging in the same direction."

The background area is assumed by our eyes to be fixed because of its size. This is also the explanation for the moving picture of the colorful dog with the purple tail, which was seen in the dark storage room at the beginning of the journey.

The following illustration shows that the combination of the two colors is important for the apparent vibration capacity. If you shake Fig. 5.22 slightly back and forth in a darkened room, the already known combinations cyan/magenta and red/green move best. Yellow and blue, on the other hand, are very stable.

5.4.7 Blue is a Very Special Color

The recognition of the color blue differs in some points quite substantially from the red and green color perception—the reason for this lies in our evolution. Gene analyses have revealed (Nathans et al. 1986) that the "invention" of blue color vision must be much older than the perception of the other colors.

The genes for red and green vision are very similar, but very different from those responsible for blue vision. From this it can be concluded that the green and red cones have a common ancestry.

Fig. 5.22 Shaking test: Which square is solid, and which is loose?

This different genesis of the cones is reflected in a whole series of phenomena. Let us first consider the frequency and distribution of the different cone types on the retina. The blue cones are very weakly represented in terms of quantity. In the fovea, the area of the sharpest cone vision, the proportion of blue cones is only 3–4%; in the area outside the fovea, however, it is significantly higher.

This different distribution of the cones inside and outside the fovea is the reason for another color phenomenon: the color of an object changes depending on the angle from which you look at it. A small color dot ●, which—viewed from the corner of the eye—appears blue-green, looks clearly green if you fix it clearly. So: It loses its blue tone when its light falls into the less "blue armed" fovea.

By the way, the other photoreceptors are also spread over the retina at different distances, as can be seen in Fig. 5.23. The rods are most widespread in outdoor areas, but they are virtually absent from the fovea itself.

The blue cones are very similar in their distribution to the rods. Their boundary line (the line on which the color blue is just identified) is only slightly narrower than

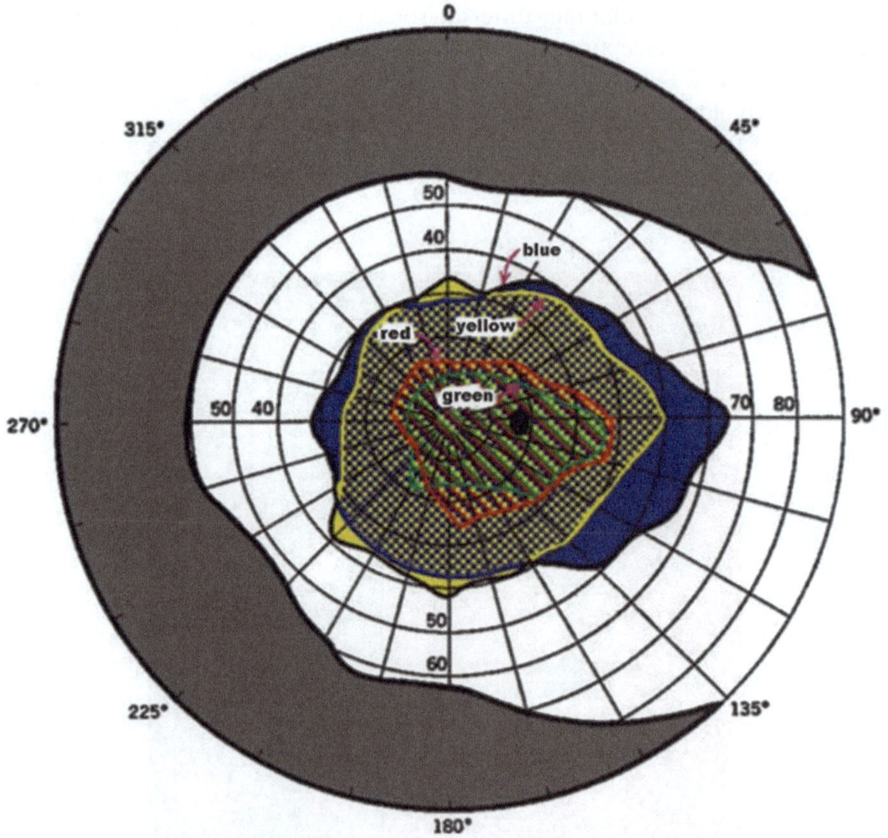

Fig. 5.23 The field of view of the right eye with the lines for the limits of recognition of a color

that of the rods. Now the circle becomes narrower and narrower towards the center of the field of view: the red cones are a little farther from the center than the green cones, which are exclusively located in the center.

This different spatial distribution of cones on the retina becomes clear when, for example, a small green or red area is slowly pushed from the side into the field of vision. At first you will only perceive the movement of a colorless surface. Only when the surface is far enough in the center of the field of view it does appear colored!

By the way, our retina performs another amazing feat when seeing the color yellow: the perception of yellow is still possible far beyond the border lines for red and green vision and in some places even exceeds the border line for blue vision—even though yellow is only perceived when a neighboring red and green photoreceptor simultaneously respond!

The explanation for the preferential recognition of yellow: In the outer areas of the retina, the photoreceptors work in interconnections, i.e., they have joined together to form emergency bandages or "neighborhood aids" and now only work together. The probability of a "yellow" neighborhood association—consisting of the combination of a red and a green photoreceptor—is much higher than the probability of the neighborhood of two similar types of cones, which are necessary for the knowledge of green or red.

In addition to the different distribution on the retina, the blue vision has another clear characteristic difference to the other colors: Thus, for example, a defective vision, as we learned with the red-green blindness, occurs with the color blue almost not at all.

Fortunately, nature progresses more slowly and gently than the computer industry in its development phases, so that even a very old model like the blue cones is still in use after such a long time. Nature even accepts a disadvantage: blue cones are slower in absorbing and processing stimuli than their red and green counterparts.

The slowness of the blue signal processing becomes clear when looking at a blue flashing light. Even at low flashing frequencies, it is no longer possible to resolve the individual flashing phases, while the same is still possible with green or red flashing lights.

Similarly, the slow processing of blue color can be detected with the help of a pendulum: Attach an eraser to a rope and let it swing back and forth in front of you. Look at this pendulum with a blue filter in front of one eye while looking forward with the other eye as normal. Suddenly you perceive an elliptical spatial movement instead of the left-right movement! The movement of the pendulum now seems to go from the left-right plane also forward and backward! The reason for this is the time delay of the "blue eye" at the entrance of the visual stimuli into the perceptual apparatus. You can read a more detailed explanation of this colored *Pulfrich phenomenon* in the forthcoming seventh journey of this book.

5.5 By the Sea

After the earthquake, the bus journey continues in the direction of the sea. After hilly landscapes, meadows and butterflies, dune landscapes, dikes and seagulls appear under a wonderful blue sky in the window.

Blue? Why is the sky blue? And why does the sun turn red at the beginning sunset and not, for example, green or yellow?

5.5.1 Why Is the Sky Blue?

"I don't think it's so important that you always know everything," says the bus driver calmly when some travelers ask him about it. "It's more a matter of enjoying it. Probably this is a physical effect, but I can't explain it exactly either. Neither can I explain why I'm still doing this badly paid job."

Since you, as a reader of this book, probably do not want to be satisfied with this succinct answer, we will try to give you a more detailed explanation. To this end, we first take a closer look at the path of white sunlight, which consists of all the different wavelengths of the spectrum, as it enters the Earth's atmosphere. The atmosphere consists of various gases, such as oxygen or nitrogen; these gases in turn consist of a large number of very small particles, the *molecules*. Sooner or later our ray of light will hit one of these molecules. What happens now in this collision?

The English physicist Lord Rayleigh calculated this impact process mathematically accurately as early as 1899. He found out that the molecules of the air *scatter* the light. This is the *Rayleigh scattering*, named after him.

As soon as a ray of sunlight comes close to an air molecule, it stimulates it to vibrate. Like a radio antenna, the vibrating molecule emits a new beam of light of the same wavelength in a certain direction. Then everything is the same again, except that the light has been scattered in one direction. Lord Rayleigh could show that there is a connection between the wavelength of light and the intensity of scattering: The smaller the wavelength, the stronger the scattering. Blue light is therefore scattered much more strongly than red light! From this important insight we can draw all relevant conclusions regarding the coloration of the sky and the sun.

All the indirect sunlight that reaches our eyes is scattered light. And since, as we already know, the visible spectrum is best scattered in blue, the sky appears blue to us! This is not a pure blue, since other color components are also present in the scattered light—albeit to a lesser extent.

And what about the sun itself? On its way through the atmosphere, the Rayleigh scattering causes the solar ray to lose more and more short-wave radiation. The sun in the midday sky therefore appears to us in yellow light, since the blue portion on this beam path is filtered out as far as possible.

The direct sun rays at sunrise and sunset have a significantly longer way through the atmosphere behind them. Instead of passing through the atmosphere vertically,

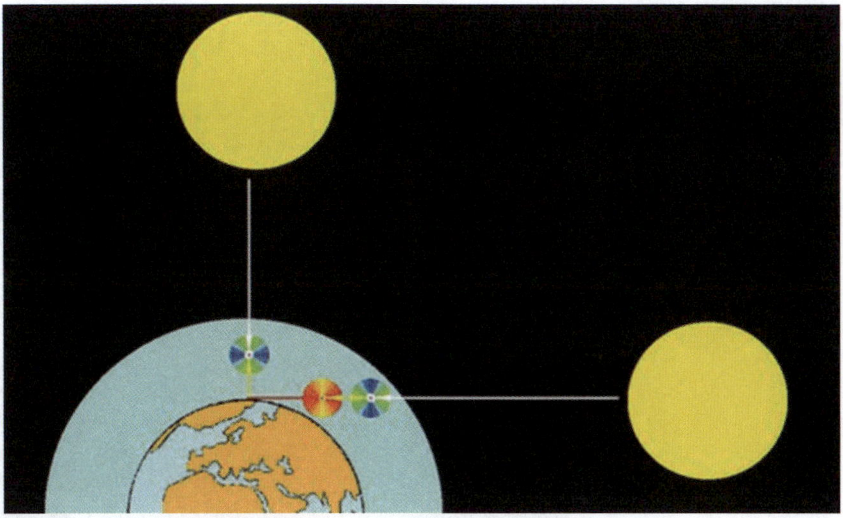

Fig. 5.24 The reason for the blue of the sky, the yellow of the midday sun, and the red of the setting sun is the Rayleigh scattering

i.e., by the shortest route, as they do at noon, they now hit the atmosphere at an extremely oblique angle. Therefore, as you can see in Fig. 5.24, their path through the atmosphere is much longer. This distance is so long that besides the blue, the green tones also are partly scattered away. Therefore only the red light remains to the setting sun, which finally reaches our eyes!

Other factors that influence the coloring of direct and indirect sunlight are the scattering of ozone or larger molecules in the atmosphere, suspended particles, or water droplets.

Other scattering mechanisms more complicated than the Rayleigh scattering also play a role here, such as the *Mie scattering*, which no longer allows blue light to be scattered preferentially. Normally, however, this additional influence on the sky colors is very small. Only at a very high concentration of such larger particles in the atmosphere is their influence noticeable: If, for example, when there is fog and there is a lot of water in the air, these scattering processes result in a cloudy white. The presence of even larger particles in the air also results in color changes. For example, floating particles from volcanic eruptions cause very intense sunsets even years later.

A further natural phenomenon, as fantastic as it is rare, is the *green flash*, which is visible for a short time on clear days shortly after sunset under certain conditions (sun coloring as little as possible red). A very nice example is shown in Fig. 5.25.

An equally beautiful description of the whole thing comes from Jules Verne in his book *Le Rayon vert* (*The Green Ray*) (1882).

> *Have you ever seen the sun set on the horizon? – Sure! – Did you follow it until the top of its disc just touched the horizon and wanted to dive down? – Very probably so. – But did you notice the phenomenon that occurs at the last ray of sunlight when the sky is clear and fog-*

Fig. 5.25 A green flash—recorded in 2001 shortly after a sunset in Madagascar over the Mozambique Channel. (Courtesy of Vic Winter)

less? – Perhaps not. – Well, the next time you have the opportunity to observe it again (it is very rare), be sure that it is not a red ray that you will see, but a green ray, beautifully green, of a green that no painter can get on his palette, a green that nature has never produced anywhere else, neither in the variety of colors of the plants nor in the color of the clearest seas! If there is a green in paradise, it cannot be anything other than this green, the true green of hope.

In this love story Jules Verne describes the long hunt of the beautiful Helena Campbell for the green flash. She won't get married unless she sees the green flash. According to an old legend, someone who has once seen this wonderful green light can never again be mistaken in love matters. The explanation for this wonderful effect after the actual sunset is based on the combination of two effects already known to us: the Rayleigh scattering and the refraction of light in the atmosphere.

The light shortly after the actual sunset is caused by the refraction of light, which we have already observed using Newton's prism. The atmosphere here has the same function as the glass in the prism. Depending on the wavelength, the light of the sun is refracted to different degrees: The long-wave components, i.e., above all red and yellow, are the least distracted and can no longer be seen geometrically by the observer due to the low position of the sun. The first candidate would therefore be blue, followed by green with some distance. But because of the Rayleigh scattering, blue does not reach us—so that the green remains!

5.5.2 The Color Contrast Enhancement

Meanwhile the bus has arrived at the sea, just in time for coffee.

Our tour group looks out at the sea and observes the color differences between sky and sea. As in Fig. 5.26, a clear boundary line between sea and sky can be seen on the horizon. In reality, the sea and the sky are not colored so differently. You can easily convince yourself of this by placing one finger across the horizon line in Fig. 5.26. The slightly darker sea suddenly seems to become much darker shortly before the horizon and the overall slightly brighter sky suddenly appears much brighter shortly before the transition of sky to sea.

With this we have made another new discovery: The brightness contrast between two surfaces increases even if they are colored! This brightness contrast enhancement of the colors is based on the same mechanisms as the black-and-white contrast, which we already know from the journey to figure and form.

This effect of increasing the edge contrast becomes very clear when you take a closer look at the stripe pattern of different blue/red tones, as shown in Fig. 5.27.

It gives the impression that the stripes, which have a uniform color, get a completely different brightness at the edges. They seem clearly bluer at the bottom and

Fig. 5.26 The view from the Beach Café: On the horizon the dark sea seems to become even darker and the slightly brighter sky even brighter. The brightness contrast of the colors increases and, despite only minimal differences in brightness, the horizon line between sea and sky is clearly visible

Fig. 5.27 Edge contrast enhancement: The uniformly colored horizontal stripes appear differently colored by the contrast enhancement, with the brightness in each strip decreasing from bottom to top

clearly redder at the top! This phenomenon is similar to the *Mach* black-and-white *stripes* you've seen before.

This observed color contrast enhancement is a very important tool for our daily vision. Only this technique—also called *simultaneous color contrast*—enables us to distinguish between objects that appear similar in color to each other.

This technique is called *simultaneous* because the colors have to be observed simultaneously and their common effect is perceived. This once again shows that colors are relative: Depending on the color environment in which they are embedded, the result is an apparently different coloring.

The apparent coloration is shifted in the direction of the complementary color of the surrounding area. Therefore in Fig. 5.28 the light-blue areas appear more blue in the upper section and more red in the lower section. In the case without separation (Fig. 5.28 right), our perceptual system produces a continuous color gradient—the effect of edge contrast enhancement is thereby extinguished.

A fascinating new example of color contrast enhancement comes from Akiyoshi Kitaoka. Please look at Fig. 5.29. Surely you see two spirals in cyan and green.

But in reality both have exactly the same turquoise green color! If you take a closer look at the picture, you can see why: one spiral is overlaid by red crossbars—this causes a shift of the color impression toward the complementary color green, according to the color contrast enhancement. The other spiral is superimposed by magenta-colored crossbars—which leads to a color impression in the direction of cyan.

These simultaneous color contrasts can be explained in a very similar way to the brightness contrast: The excitation stimuli of the different visual cells are compared to each other in the visual center in such a way that the values of neighboring visual cells influence each other. This spatial combination of the individual color stimuli

Fig. 5.28 The simultaneous color contrast: Consider the different color effect of the two identical bluish areas above and below. If the surfaces meet as in the right part of the picture, the effect disappears!

Fig. 5.29 The two apparently different colored spirals (green and cyan) actually have identical colors. (Courtesy of Akiyoshi Kitaoka)

takes place in the layer of so-called color antagonistic cells. This summary of data in the downstream cells takes place using a very simple trick that has been known for a very long time.

5.5.3 Hering's Anti-Color Theory

As early as 1878, the physiologist Ewald Hering developed a theory of color vision named after him: *Hering's anti-color theory*. This theory was developed from observations of afterimages, color mixtures, and color-fail-sighted test subjects.

He compared different colors with each other and came to the intuitive conviction that there are four basic colors: Blue, green, yellow, and red. Hering noticed that these four colors have a very special symmetry. Take, for example, the color blue: blue can be additively mixed very well with green and red—the result is blue-green and violet. On the other hand, a mixture of blue and yellow does not result in a color of its own, but rather in white. Hering concluded from this that blue and yellow must be opponents in the visual center—just like red and green.

According to Hering it is sufficient to know about the respective status of the "competition" between blue and yellow and between red and green in order to recognize the colors. Hering suspected that these two duels would take place independently of each other. He concluded this, among other things, from the observation that people who cannot see green cannot see red, either. People who are color-blind to blue can't see yellow.

In fact, exactly these opposite processes take place independently of each other in the visual tract! In addition to the red-green system and the blue-yellow system, there is also a third system: the light-dark system.

These systems can be imagined as three weighing pans, as shown in Fig. 5.30. Each of these scales is responsible for a specific combination of neighboring photoreceptors on the retina and is realized by the horizontal cells and bipolar cells known from the first voyage.

If, for example, a message comes from a blue photoreceptor cell, it arrives on the far left side of the scale of the blue-yellow process and deflects it in favor of blue. In this case, an incoming green arrives exactly in the middle of the blue-yellow scale, so it has no influence on the "judgement." In the red green scale, however, it arrives at the right end and influences this competition in favor of green.

The entire color part for a retinal area comes from the position of these three scales, i.e., the individual judgments about the three-color competitions. The judgement is passed on to the following cells in the form of a stimulus. The more a weighing pan tilts to the right, the more positive this stimulus is; the more it tilts to the left, the more negative the stimulus is; in the event of a tie, the result is zero.

Each incoming color has a different influence on the three processes and their result stimuli. Yellow-green, for example, reinforces the blue-yellow process and inhibits the red-green process. It is also said that the incoming colors have an *activatory* or *inhibitory* effect on the individual processes.

The phenomenon of color contrast enhancement, as in black-and-white vision, results again from the ingenious spatial combination of the neighboring cells. The spatial grouping takes into account the position of the respective photoreceptor in the cell area. If it lies somewhere in the normal, central area, its visual stimulus is received by the "weighing pan cells." If, however, the photoreceptor lies at the edge

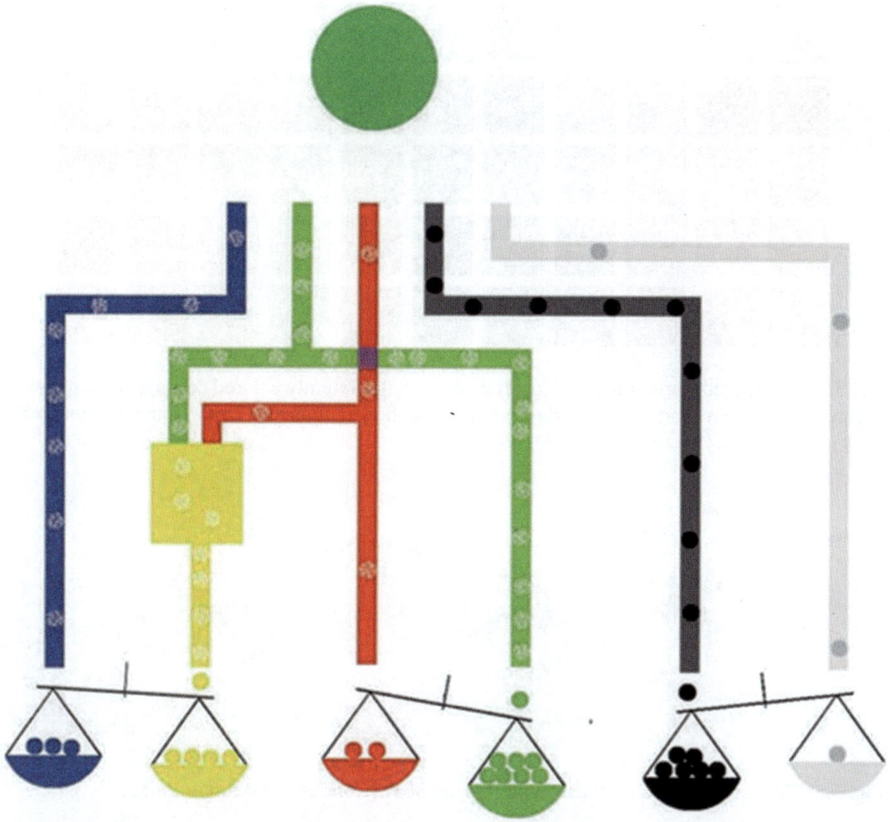

Fig. 5.30 Hering's anti-color theory: There are three downstream processes for a certain retinal area, each of which comes to a "judgement" independently of the other. The color impression only results from the position of all three scales together

of the range, its influence on the respective weighing pan is reversed, i.e., a color that normally deflects the scale to the left now deflects it to the right! In other words: In these visual cells located in the outer area, inhibitory colors become activatory colors, and inhibitory colors become activatory colors.

The simultaneous color contrast and the color contrast enhancement are produced with the help of this interconnection technique! How functional the color contrast enhancement actually is can best be illustrated by a colored Hermann grid (see Fig. 5.31). The different colors at the white intersections are created by combining the color contrasts of the colored squares involved.

By the way, Hering's theory of anti-colors can be used to explain colors such as brown or olive green, which are neither found in the rainbow spectrum nor can they be mixed together additively from the colors of the rainbow (for example, with the aid of slide projectors). These colors are only created by special color contrasts. For example, the impression of brown is only created when a yellow or orange light spot is surrounded by an average amount of brighter light.

Fig. 5.31 A colored Hermann grid: Not really existing, differently colored dots appear at the white intersections. The reason for this is the formation of the simultaneous color contrast to the differently colored adjacent areas

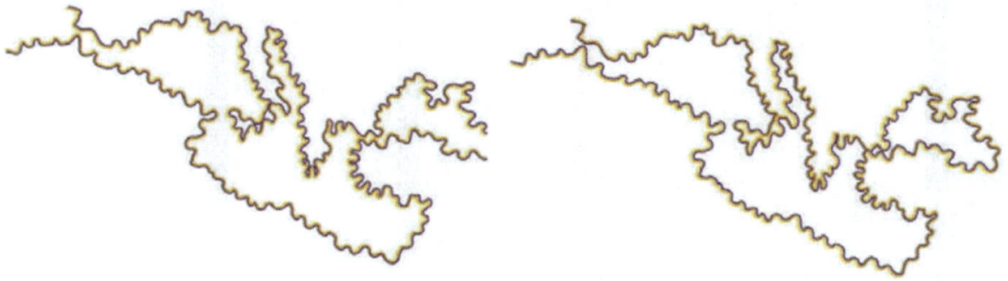

Fig. 5.32 Yellow intestine or blue Mediterranean Sea? (Courtesy of Baingio Pinna)

5.5.4 *The Watercolor Effect*

In Fig. 5.32 you can observe a wonderful effect that shows the uncanny power of colors with the simplest means. The *watercolor effect*, discovered by Baingio Pinna (Pinna 1987; Pinna et al. 2001, 2003), shows that a clear color spread can be achieved with the aid of edge colors alone.

In Fig. 5.32 on the left, for example, you can see an abstract yellowish figure reminiscent of an intestine. In reality, however, the color of the figure is pure white! The apparent coloring is only due to the yellowish coloring of the edges. Even more amazing is the impact of this effect when you look at the right part of Fig. 5.32, where only the arrangement of the yellow and blue edges is reversed. Now the intestine-like figure appears only very vaguely in the background and slightly bluish, while the outer part of the picture appears yellowish and in the foreground. The mere coloring of the edges thus determines the color impression of the entire adjacent surface and influences the very strong design law of the good form—with the consequence that we now clearly recognize the intestine as the Mediterranean!

Remarkably, some historical world maps use the watercolor effect. For cost reasons, the cartographers presumably refrained from completely coloring individual

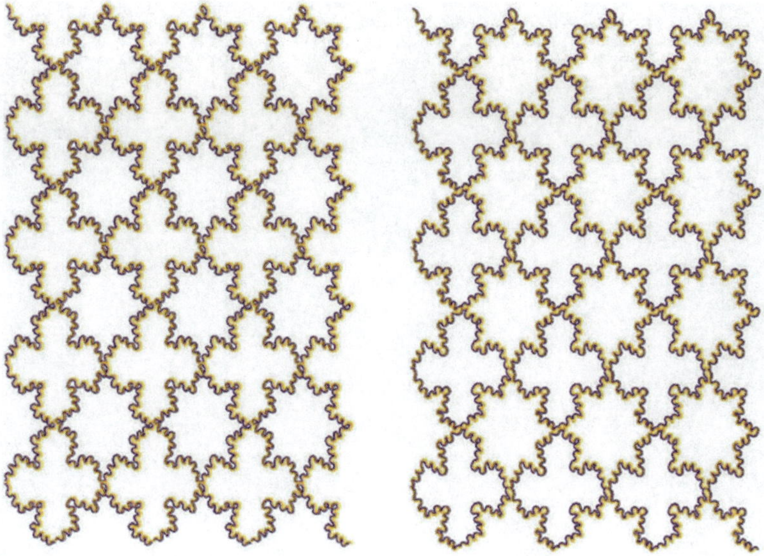

Fig. 5.33 Crosses or stars? (Courtesy of Baingio Pinna)

countries and limited themselves to drawing the borders with lines of different colors on both sides.

Another example of this phenomenal effect can be seen in Fig. 5.33. On the left and on the right the identical basic patterns with different edge colors are depicted again. Nevertheless, the Watercolor effect causes a completely different coloring and figure: On the left, yellow crosses, and on the right, yellow stars, appear as figures in the foreground. Just how strong and far-reaching the color effect is you can see from the fact that the entire middle area between the two images appears to be colored in yellow.

In this clarity, the watercolor effect is only achieved for very specific combinations of color, brightness, and background. It looks as if an effect occurs only when an edge is not very different from the background in its brightness and color. Only in this case (like the yellow in Figs. 5.32 and 5.33 on a white background) does a clear coloration of the involved area appear to be visible. The other edge, on the other hand, should be clearly different in its brightness and color.

The influence of different background brightness levels on Fig. 5.33 can be seen in Fig. 5.34. The identical pattern of Fig. 5.33 can be seen in front of a background with brightness decreasing from bottom to top. In the lower image area the impression of Fig. 5.33 can be exactly observed: Yellow crosses can be seen on the left and yellow stars on the right. But the farther up we look, the more the yellow coloration disappears and the effect even turns around completely: Blue stars can be seen on the top left and blue crosses on the right!

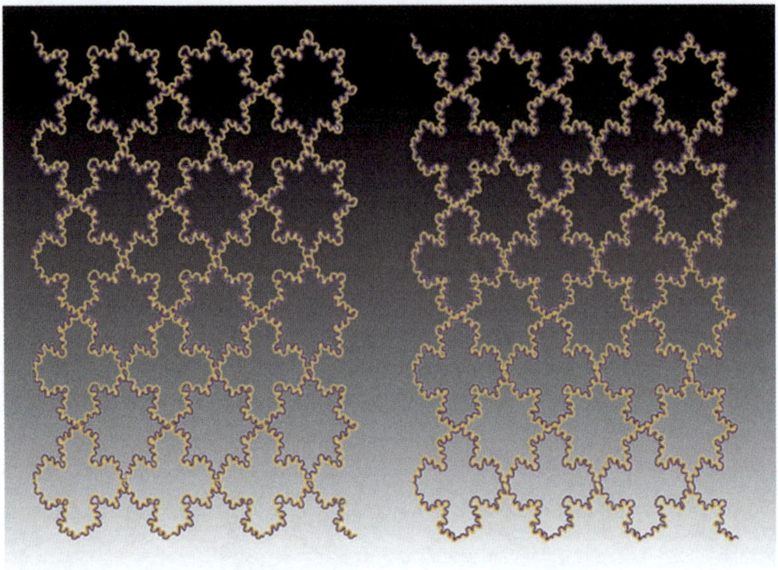

Fig. 5.34 Crosses and stars dependent from the background (Courtesy of Baingio Pinna)

Thus we have arrived at the end of the journey into the fantastic world of color perception. In the meantime—after a fascinating sunset, it has become dark. Most fellow travelers are so tired from the day's experiences that they immediately fall into a deep sleep.

They dream of meadows with butterflies, a dog wobbling its purple tail, earthquakes, sunsets, and grey cats. You now know that cats not only appear grey at night, but also fuzzy and slow. And that colors possess a magical power that can simply "laugh away" the laws of seeing.

Chapter 6
The Sixth Journey: Spatial Vision

This journey through the 3D world takes us to a land of almost boundless freedom. By combining the visual impressions of both eyes, many new effects of seeing can be achieved and old familiar ones rediscovered in a new form.

We encounter visionary stereo images with mysterious depth layers and environmental impressions of three-dimensional dimensions. The 120 million rods and the 6 million cones of our retina are allowed to work at full capacity in every respect. Let us dive into the wonderful world of three-dimensional seeing and amazement, and let us be enchanted!

6.1 Before Departure

It would certainly not have been too difficult for nature to repeat an invention once it had been made, like the human eye, any number of times. This is made clear by frightening recent research results in which fruit flies with a multitude of eyes were bred in a wide variety of body parts through genetic manipulation.

6.1.1 Why Do People Have Two Eyes?

It's the most natural thing in the world: people have two eyes. But why do we have exactly two eyes and not just one or three or more eyes? Probably you would have been happy about a third eye, for example at the back of your head. Or over one eye

T. Ditzinger, *Illusions of Seeing*, https://doi.org/10.1007/978-3-030-63635-7_6

on each finger so you can look around every corner. Or about an eye that is active at night when the other two eyes are resting.

You can look forward to a joyride into the wonder world of three-dimensional vision with many junctions, dead-ends, and side roads. But dead-ends and side roads are known to be the attraction of joyrides.

6.1.2 The Eyes

At night our eyes rest from their day work. But this happens in a very strange way: by working like wild! Yes, you've read that correctly: In the famous *REM (rapid eye movement)* phases, the eyes move back and forth completely uncontrollably and quickly. These REM phases have a length of 15–30 min and run during one night at increasingly shorter intervals of about 90 min at the beginning. If people are awakened during these phases with rapid eye movements, they can reconstruct their dreams very accurately. This is an indication that the REM phases are associated with a very intense dream activity. In the REM phases both strong activities of the mind and the visual system take place. The assumption suggests itself that these activities of eye and brain show a connection in the dream state.

Why does man need this "night work" to recover from "day work," and what is the difference between these two forms of work?

During the day, our brain and visual system have almost only meaningful tasks to perform. These are constantly predetermined by external stimuli or determined by our consciousness. Night work, on the other hand, looks quite different. As soon as the "boss" (the consciousness) is out of the house or rather has fallen asleep, the mice are dancing on the table! This is no different from working life. Thus, many a stressed employee suddenly becomes a "new" person as soon as he or she is able to act in a self-determined manner after the end of work. In the evenings, voluntary jogging or forest walks are made, or other unusual physical or mental activities are performed. The "night work" is completely decoupled from constraints and external determination. It is the same in the dream phase: the thought activity connected by our consciousness to chains of association is now completely decoupled and free. Likewise, the eye movements of the two individual eyes are completely decoupled from each other.

6.1.3 Coupled and Decoupled Eyes

You can easily convince yourself that this is by no means the case during the day with the following experiment:

Close one eye and place one finger on the eyelid. If you look now with the opened eye in a certain direction, you will feel that the other eye is carried along automatically. Thus, there is a coupling of the visual axes of the two eyes. You can even check this with two closed eyes: Keep both eyes closed and then register the movement of the eyes with a finger on each eyelid! Then draw your attention to a few things that you know are in a certain direction. You can already feel through your fingers that the two eyes are simultaneously turning toward these things. Here, too, the movement of the visual axes is coupled together. This ability to couple the lines of sight of the eyes is the basic prerequisite for *stereo vision*.

Experiments in which the perceptual impressions are artificially altered show how important it is to coordinate the two eyes with each other. The test subjects are fitted with various special glasses; these consist of mirrors, colored lenses, and prisms and, depending on the design, can interchange the impression of top and bottom, left and right and colors, and can even make straight stretches appear curved. This naturally creates a hopeless mess of the senses. Interestingly, long-term experiments have shown that our perceptual apparatus is very adaptable and that we are able to find our way in this new environment after about a week.

All people who are blind in one eye cannot naturally achieve a coupling of the two eyes and thus miss the basic prerequisite of stereo vision. In addition, there are about 4–6% of the normal-sighted population who are not able to adjust both their eyes to the same observation point and are not even aware of this defect. The proportion of people who cannot see three-dimensionally is estimated at up to 15%.

The most frequent cause of the lack of such depth perception is *strabismus*. Strabismus can have various triggers and occur as both inward and outward strabismus. This inability to place the eyes parallel can occur as early as childhood, due to, for example, eye muscle abnormalities, defects in the brain stem (which controls eye movements), or external injuries.

In childhood, our visual system develops various strategies to compensate for this defect. One of them is *alternating strabismus*. The two eyes alternate at a distance of approximately 1 s when focusing. During this time, seeing through the other eye is suppressed. A second strategy of the visual system to compensate for the cross-eye defect is the constant *suppression of one eye*. This, however, worsens the vision, and in extreme cases the eye can even go blind. Very often, however, people squint only in a weakened form and are therefore unaware of their inability to see spatially.

If the strabismus does not occur until after childhood, our visual system is no longer able to compensate for this defect. If, for example, a boxer cannot keep his eyes parallel after a fight injury, he will inevitably see double images. Each eye now delivers its own visual information separately and the brain is no longer able to combine this different information into a three-dimensional image.

You can easily try out for yourself how it feels to someone who suddenly squints due to an injury: First, fix an object in your field of vision with both eyes. Then press carefully from the side or from below on one of your eyes. Immediately the view

Fig. 6.1 Normal vision and strabismus

becomes somewhat blurred and glassy, until you finally see double images, which of course are no longer three-dimensional!

If you have a little more time for a self-experiment, you can also reach this state through a strong influence of alcohol. Its effect on the eye position can be seen in Fig. 6.1. However, we do not assume any guarantee for any other undesirable side effects.

In addition to strabismus, there is another reason why some people are not able to merge the two individual images of the eyes into one impression of depth—namely, different visual acuity of the eyes. If the lenses used to compensate for this visual weakness differ by more than three diopters, the person concerned finds it very difficult to perceive depth.

The two very different lenses produce retinal images that are very different in size and therefore can no longer be matched by our visual system.

One possibility for decoupled vision of the two individual eye images is also the "first look" after waking up in the morning, which can also be used as an excuse to remain lying down for a longer period of time:

Take your time getting up! First open your eyelids very gently and carefully. Stare into the air without changing the direction of both eyes. You will now clearly see two images that you can hold still for as long as you like, as long as you keep your line of sight stable. But once your eyes are parallel, you won't be able to do this trick again today—unless you extend the experiment by going back to sleep.

6.1.4 Three-Dimensional Environmental Impressions

With the recognition of the two uncoupled individual images of the eyes, we have come very close to the door to understanding stereo vision. We live in a three-dimensional world, are ourselves three-dimensionally built and, in order to find our way in the environment, have to perceive these three-dimensional environmental impressions accordingly. This is physically very difficult to do with a single eye. This is because each eye reproduces the incoming information in two dimensions on its retina, like on a kind of canvas.

The physicist Hermann von Helmholtz (1821–1894), who met us on our fifth voyage, once said: "I would throw out an optician who would bring me an instrument like the human eye!"

Nevertheless, he could never rebuild a camera with the same qualities. In fact, the shape of the eye and the centering of the imaging surfaces (cornea and lens) are even worse than with a cheap camera—not to mention the imaging errors. In addition, a sharp image on the retina is severely impaired because, for nutritional and switching reasons, other cell layers lie in front of the light-sensitive cells of the retina, as we already saw in the first journey.

However, all these difficulties are largely corrected by sophisticated control mechanisms in the visual center. In addition, the human eye far surpasses any camera in the area of light intensity perception: The eye can perceive differences in light intensity of no less than 15 powers of 10 (!) in the brightness range between nighttime darkness and the greatest daytime brightness.

The result on the retina of a single eye can be simplified like a photograph: colored, pin sharp, and high resolution, but flat!

There is no depth information in the retinal image—you can easily check this by keeping one eye closed. In the same way, the other eye naturally produces a two-dimensional image of its field of vision on its retina. However, this is shifted by about 6½ cm from the first eye!

This shift is the decisive trick of nature through which we are enabled to see spatially. The visual axes of the eyes, and thus their two inner canvases, are inevitably tilted against each other! Therefore the inner total canvas, which one can imagine composed of the two retinas, is no longer only two-dimensional, but three-dimensional! It is not only two independent single images that are reproduced on the two two-dimensional screens. Rather, our brain is able to calculate a three-dimensional perceptual impression from these two images!

The visual center even has various possibilities for this: On the one hand, depth can be determined by observing the angular position of the visual axes of the two eyes when both fixate on a certain object: the so-called *convergence*.

The second possibility is to observe the comparison of the differences between the two images in their width deviation: the *transverse disparity* of the two images.

6.1.5 Depth Determination Through Convergence

Determining the depth of an object from the convergence is very simple.

This is why people have long been using this idea of nature, for example in astronomy, to measure cosmic distances. Understandably, optical distance measurement methods such as measuring the curvature of the lens when in focus, fail completely at astronomical distances.

The distance of an astronomical object such as Jupiter, on the other hand, can be measured very easily if you aim at it with two telescopes. All you have to do is compare the observation angle of the two telescopes. From the difference, the

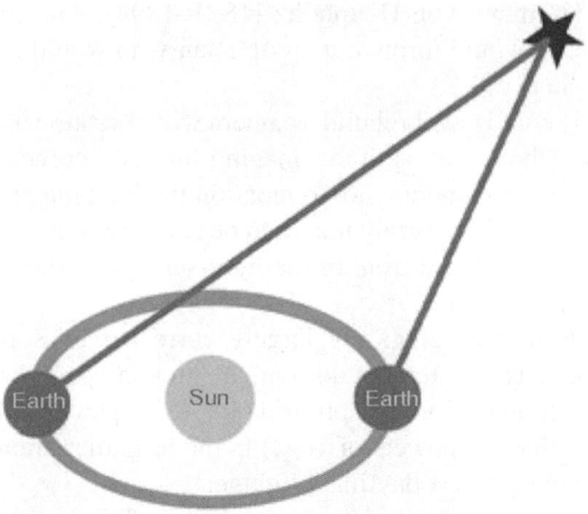

Fig. 6.2 Distance measurement in astronomy: the position of an astronomical object is targeted with two telescopes at different locations. The telescope axes are at different angles to each other. From this parallax angle the distance of the observed object can be determined. If one uses the rotation of the earth around the sun, and if one compares the telescope parallax of the star at the most distant points in the distance of half a year, then distances up to approximately 100 light years can be reliably determined

parallax, the distance of Jupiter can be determined with the help of the laws of ray optics. The measurement error becomes smaller the farther the observation distance of the two telescopes.

This is why it makes sense to take advantage of the Earth's motion around the sun. If the same object as in Fig. 6.2 is observed again at a distance of half a year, the distance of even more distant objects (such as "near" fixed stars at a distance of up to 100 light years) can be determined relatively accurately.

Distance measurement by means of eye convergence could work in a similar way: The two eyes fixate on any object in their periphery—it is also said that their visual axes converge on the object. The distance of the object can then be deduced from the parallax angle. Nature would therefore only need to "feel" the position of the eyes or monitor the muscular activities necessary to achieve convergence in order to measure distance by means of convergence.

Surprisingly, however, nature offers itself the great luxury of not making further use of this distance-measuring method! No sensors have been discovered in the eye, at least so far, that exploit the property of binocular convergence. This generosity with its resources can only be explained by the fact that nature has succeeded in a far more ingenious, but also more complicated invention, which includes the convergence measurement as a special case: the transverse disparity!

6.1.6 *Depth Determination by Transverse Disparity*

In contrast to astronomical distance estimation, man has to estimate the depth of a fundamentally different environment. The human environment does not "only" consist of individual light points, which are to be examined for their different depths. Rather, the human visual apparatus must solve a far more complex task: Our environment consists of entire composite bodies that can protrude into the most diverse layers of depth. Ideally, the eyes should not only be able to examine a single point of light for its depth, but should also be able to assign the correct depth to their entire field of vision at a stroke.

This miracle is possible all by itself—and even incredibly accurately—by comparing the two images on the retinas of the two eyes. This means that the overall result of three-dimensional vision is far more than just the sum of the two-dimensional individual image parts! Rather, it is a completely new, individual visual experience. In the following, we will examine in more detail how the combination of the individual images takes place. The itinerary leads us through the 3D-land to the "Index Finger path."

6.2 The Index Finger Path

6.2.1 *A Vertical Index Finger*

Hold an index finger vertically in front of the tip of your nose at a distance of approximately 20 cm as shown in Fig. 6.3. Now close your left eye and observe your index finger with your right eye. Remember exactly what can be seen behind your index finger. Then observe the same scene with your left eye only.

The index finger has clearly moved to the right compared to the background! Our two eyes thus provide a very different impression of the same scene. The closer the objects are to us, the stronger is the horizontal difference of their embedding in the background.

Pull your index finger a little closer to you and repeat the experiment. The horizontal distance is now much larger! This different width deviation in the overall scenery is called *transverse disparation* and is the basis of our spatial depth vision.

Fig. 6.3 Index finger in front of the nose

The human visual apparatus can combine the two different double images on the basis of their transverse disparities in such a way that we obtain a completely new, three-dimensional sensory impression. Our brain has learned how to connect the incident amounts of information of both eyes so cleverly that a brilliant depth resolution is possible. It is known that each eye contains approximately 126 million photoreceptors, whose information is compressed to approximately 800,000 sensory impressions after a kind of pre-processing by the subsequent ganglion cells. This information supplied by the eyes is then compared simultaneously according to mostly very simple laws.

6.2.2 Two Vertical Index Fingers

Hold an index finger vertically in front of the tip of your nose at a distance of approximately 20 cm, as you did before. The other index finger should be placed with outstretched hand in vertical position behind the first index finger again. Now look with both eyes at the index fingers. You will certainly not be able to focus both index fingers at the same time. Our perceptual system has thus reached a limit, for it is not created for such "unnatural" extreme cases. Now focus your gaze on the closer finger and observe the rear finger from the corners of your eye—or rather the rear *fingers*! Because two index fingers are clearly visible in the depth: one on the left and one on the right—seen from the front finger. If you don't immediately see two fingers, it's because we're trimmed to ignore these double frames. If it has not worked after some time, it usually helps to move the rear index finger a hand's width up and down (Fig. 6.4).

Fig. 6.4 Two vertical index fingers

Our perceptual system behaves the same way in the opposite case: Focus your eyes on the back finger. Suddenly two fingers appear in front: one on the left and one on the right. The explanation is very simple: Our depth perception system is no longer able to fuse such extremely distant objects into a three-dimensional impression—their transverse disparity is simply too large.

Therefore, one can only see the two-dimensional double images with the index finger, which lies just outside the attention.

6.2.3 Two Horizontal Index Fingers

Try another impressive index finger game with double images, as shown in Fig. 6.5.

Hold both index fingers horizontally in front of you and let both fingertips touch each other. Imagine now the outstretched index finger from the previous experiment was still there. Now focus your view on this distance—or you can just fix the background. Suddenly you see a fleshy, sausage-like structure between your two fingers. Now slowly pull your fingers apart and watch what happens to the sausage: The sausage seems to become weightless and float freely in the space between the two fingers. If you pull your fingertips apart, they become smaller and smaller until the whole spook suddenly disappears again. Hocus-pocus fidibus!

Behind this trick is a whole new way of perceptual deception. The aim of our visual apparatus is always to achieve three-dimensional perception. This means that the two retinal images are connected as well as possible. However, our visual system does not always manage to fuse the right objects with each other. In this experiment we are misdirected in such a way that the retinal image of the left finger is

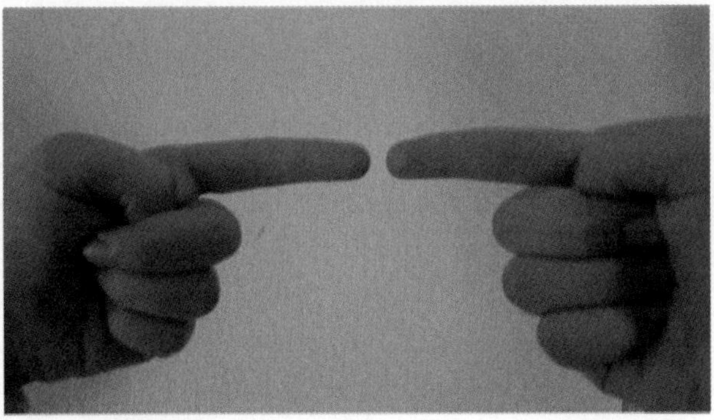

Fig. 6.5 Two horizontal index fingers

fused with that of the right finger. This is achieved by the convergence of the eye position at a distance. On the left retina the left finger appears in the center and on the right retina the right finger.

If you have succeeded in this "trick" you can be pretty sure that you won't have any difficulties with the other test images for three-dimensional perception in this book. Otherwise, you may be among the 4–6% of people who are unable to see spatially.

6.2.4 Depth Resolution due to Transverse Disparity

By being able to convert transverse disparities into a three-dimensional overall impression, humans possess a fantastic tool for correctly estimating and resolving depths. The greater the visual acuity and distance between the eyes, the better the resolution. For example, successful tennis players were often found to have greater eye distances than normal: 7 cm and more.

In addition, the greater the distance from the lens to the retina (on average 1.67 cm) and the smaller the distance between the individual photoreceptors, the better the depth resolution (in the central retinal area, the fovea, the distance between the photoreceptors is about 2 μm). From the fovea to the retinal periphery, the distance between the photoreceptors increases more and more. One can assume that an average eye can resolve minimally different visual rays up to an angle of 3.6 arcsec—that is 0.001°. This is an almost sensational value and does not lag so far behind the resolution of parallax measurements in astronomy with 0.0016 arcsec, whose "interpupillary distance" (diameter of the Earth's orbit around the Sun) is about 150 million km, which is so much greater. Thus it becomes only too understandable that nature makes no use of the relatively "rough" method of measuring convergence.

Further examples of the performance of our depth resolution capability using transverse disparity are given in Table 6.1:

You will now probably have the impression that your depth resolution is better than shown in the table. Above all, the resolution values at distances from 500 m appear to be significantly worse than we believe to know from our experience. We

Table 6.1 Depth resolution by transverse disparity

Visual distance	Resolvable depth differences
Reading distance (35 cm)	0.13 mm (breadcrumbs)
Room wall distance (3 m)	0.5 cm (picture frame)
Nearer surroundings outdoors (20 m)	22 cm (hand length)
Football stadium (100 m)	5.7 m (half football goal)
500 m	180 m

From a viewing distance of approximately 1 km, depth resolution is no longer possible this way.

will try to find out the reason for our supposedly better depth resolution in the further course of the journey.

6.3 The Random Dot Images

Look at Fig. 6.6 with the randomly distributed black and white dots—allow enough time for your perception. Suddenly you will see something new in the picture. This self-organized fantastic spatial phase transition of your perception is undoubtedly one of the most impressive experiences of seeing, especially if you had never experienced it before.

6.3.1 The Trick with the Stereo View

Image pairs such as those in Fig. 6.6 were first developed by Bela Julesz in 1960 and were the starting point for experimental research into stereo vision (cf. Julesz 1971). The two meaningless images are wonderfully merged by our depth perception into a three-dimensional spatial impression. This suddenly causes a triangle to appear in the depth, floating above the background. The principle of this trick is quite similar to the experiment with the two horizontal index fingers and the discovered floating fleshy sausage in Fig. 6.5.

Hold the book in front of you at a reading distance. It is best to look through the book now—just as if it were a glass pane. This is very easy if you focus your eyes on a much lower point, such as an index finger on the outstretched hand. Now relax your eyes and give them time to fuse the perceived double images. The left eye should look through the left random image and the right eye through the right image. This special alignment of the eyes is shown in Fig. 6.7.

To be on the safe side, you can place a cardboard box between the pictures. The images overlap in one fell swoop—and you know how the trick works.

Fig. 6.6 A simple random dot stereogram: What do you see when you look deeply?

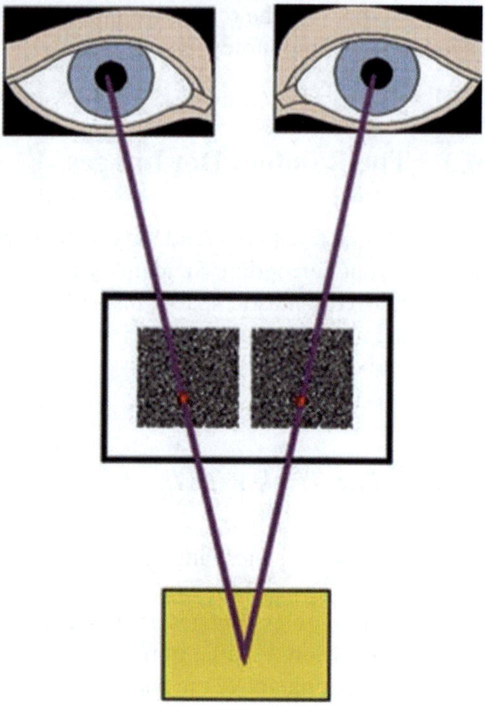

Fig. 6.7 The stereo view: With this orientation of your gaze, you can see the depth information of the random point stereograms

6.3.2 The Creation of Random Dot Stereograms

The amazing thing about random dot stereograms is that no structure can be seen in the images. What is behind these pictures and how are they made? See Fig. 6.8 for the individual steps involved in creating a random dot stereogram:

1. First, any random picture is painted or printed with the computer.
2. Now this random image is copied and pasted next to the first one.
3. Using scissors or a computer, a figure, for example a circle, is cut out of this second image and glued on again, horizontally shifted by a distance.
4. The resulting gap is filled with random dots.

The two pictures are identical in very large part, although this is not noticeable at first glance. If the two retinal images are brought into congruence at the correct depth, all the tiny random structures spatially connect at a certain depth. The prepared figure necessarily has a different depth than the background because it has a different width deviation. For example, if the figure was moved to the left, it appears in the foreground. The more it was shifted, the more it is perceived in the foreground. Conversely, the figure appears behind the base if it has been moved to the right after cutting.

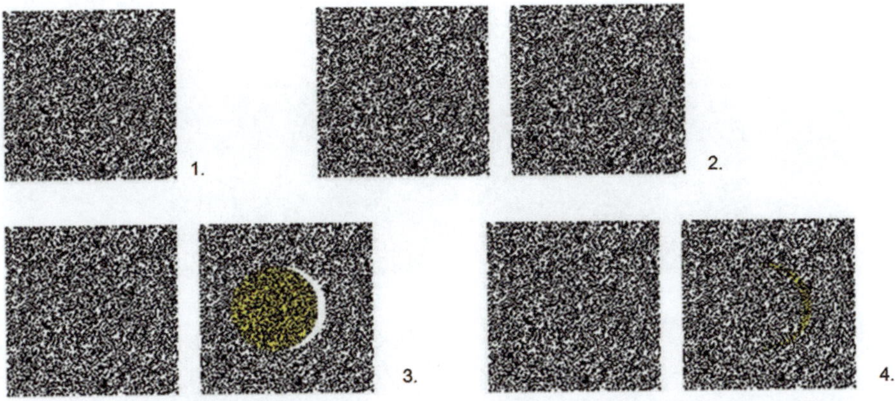

Fig. 6.8 The creation of a random dot stereogram: (1) First, any random picture is painted or printed with the computer. (2) Now copy this random image and paste it next to the first one. (3) From this second picture a figure, for example a circle, is cut out with scissors or a computer program and—shifted horizontally by a distance—glued back on. (4) The resulting gap is filled with random dots

With the help of random dot stereograms we can gain an important insight into our perceptual system: At first we recognize spatial depth and only then as next shapes and forms. It is therefore not necessary for stereoscopic vision to recognize any structures or forms in the retinal images of the individual eyes.

6.3.3 Fantastic Experiments with Spatial Perception

The depth impression of the random dot stereograms thus results solely from the effect of transverse disparity. These images are therefore an ideal way to test your stereoscopic depth-perception capability.

These figures can actually only be seen if you can see them spatially—cheating is out of the question!

Furthermore, the random point stereograms are the ideal measuring instrument to objectively check the strength of depth perception without other disturbing influences. You can experience this in Fig. 6.9. There you can see three test images to check your depth perception strength: Which figure can you still see at close reading distance? Circle, square, cross? The cross requires the least depth perception, the circle the strongest.

In addition, the Julesz random point images give us a whole fireworks of amazing new insights into the basic principles of our visual system. The next pictures will show that our brain also strives for the simplest state in three-dimensional perception and is also very inventive. Make the following eye tests:

First eye test: Look at the two stereo image pairs in Fig. 6.10. At the top you can see two horizontal bars in the foreground. In the lower picture pair, the space

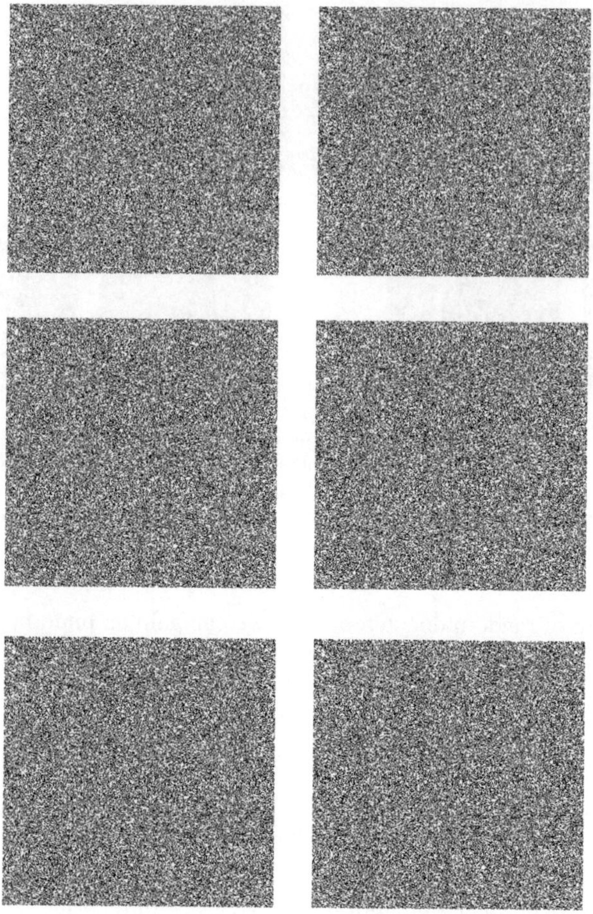

Fig. 6.9 Test images to check your depth perception strength: Which figure can you still see at close reading distance? Circle, square, cross?

between the two horizontal bars is left white, otherwise everything is the same as above. That is why our perception here has two choices. Either the two individual beams remain in the foreground or a closed large area. Observe for yourself what your depth perception does!

The decision is clearly made in favor of the simpler perception, i.e., the closed large area in the foreground.

Second eye test: In both image pairs (Fig. 6.11), a square can be perceived floating in the foreground. The square was shifted from the left to the right part of the image only minimally by one pixel to the left (transverse disparity minus 1). The pattern of the random dot picture is so cleverly prepared here that the square can lie at various altitudes. This is achieved by repeating the texture of the square in every second pixel.

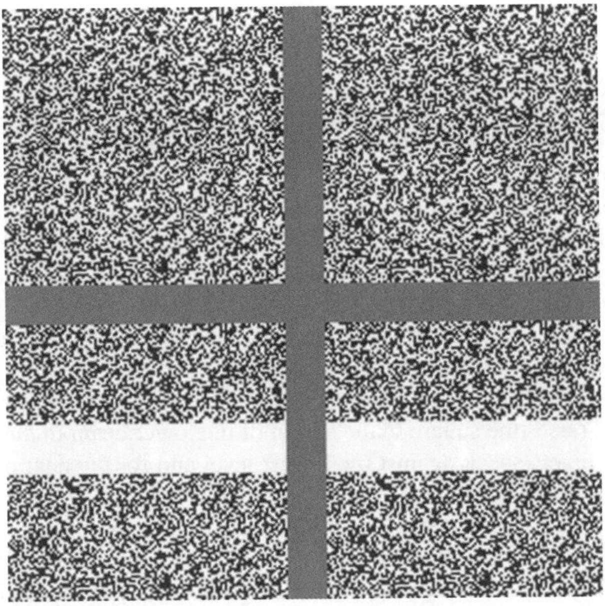

Fig. 6.10 Two beams or closed surface?

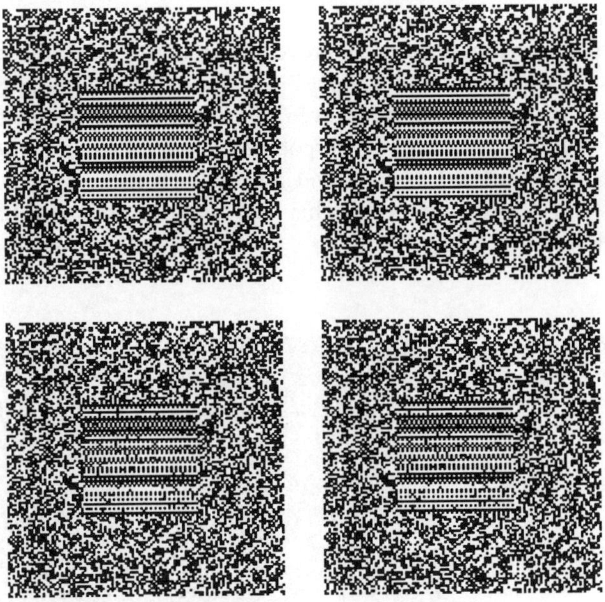

Fig. 6.11 Second eye test: the Pulling effect

The lowest possible level (disparity minus 1) is represented by the height of the horizontal bar above the square, the next level (disparity minus 3) by the bar below the square.

In the upper part of the picture, our visual apparatus has a freedom of choice with regard to the perceived height level of the square. As you can see for yourself, the decision is once again made for the simplest solution: the square is perceived at the lowest height. So it floats at the same level as the upper crossbeam. Our visual apparatus behaves as if it were a lazy weightlifter and the square made of lead. So we can hardly speak of "hovering above the background"!

In the lower part of the picture, a very small proportion (about 6%) of pixels is inserted into the ambiguous square from the upper row, which clearly indicates a disparity of minus 3, i.e., the second height level.

Again, our perception goes the easiest way. It concludes a compromise of least resistance and raises the square to the height of the lower beam in the second stage. Since the 6% pixels speak against the lowest level and for the next possibility, this depth is clearly recognized!

Third eye test: The texture (in Fig. 6.12) of the triangle is designed in such a way that half of the pixels indicate a disparity of minus 3 and the other half has no disparity at all. Do we now see a floating triangle, or is it exactly in the background? The solution is again a surprise of our perception and nature: Transparency!

With fantastic ingenuity, our perceptual system again detects the best possible solution. Since the two simplest possible solutions are not clear, the result is a completely new, fantastic compromise: the triangle floats in the foreground, but appears to be transparent!

Our perceptual system once again proves to be an ingenious inventor and at the same time a "thoughtful patent attorney." As Figs. 6.10, 6.11, and 6.12 show, the recipe for success is always the same, even when connecting the two eyes: First, the simplest option is considered! If this hypothesis is free of contradictions, it is "perceived," otherwise regarded as false and rejected. Then the next possible simple hypothesis is considered, and so on. This will continue until a workable compromise is found.

Fig. 6.12 Third eye test: Floating or lying triangle?

Our journey through the 3D world takes us now to a land of almost limitless freedom. By combining the visual impressions of the two eyes, many new phenomena of seeing can be achieved and old ones can be rediscovered in a new form.

6.3.4 Noisy Images

What happens when our two eyes are shown images whose texture is only congruent to a limited extent? To what extent can our eyes still match these images?

Let's just stay for a further look to the random dot image pairs to examine this question. See Fig. 6.13 for more information. When you recognize the upper triangle in Fig. 6.13, your perceptual apparatus has already achieved another quite astonishing performance. The right image is the same as the right one in Fig. 6.6, but additionally covered with 50% "noise": Every second pixel thus consists of information that is completely useless for depth perception.

The lower image is even covered with 90% noise. Although now only every tenth pixel provides meaningful 3D information, our perception is still able to combine these image pairs into a three-dimensional impression!

The urge of our visual system to connect the two different retinal images is therefore enormous. In the case of less structured images such as random dot images, it is quite acceptable that only a small amount of image information in the two individual images matches.

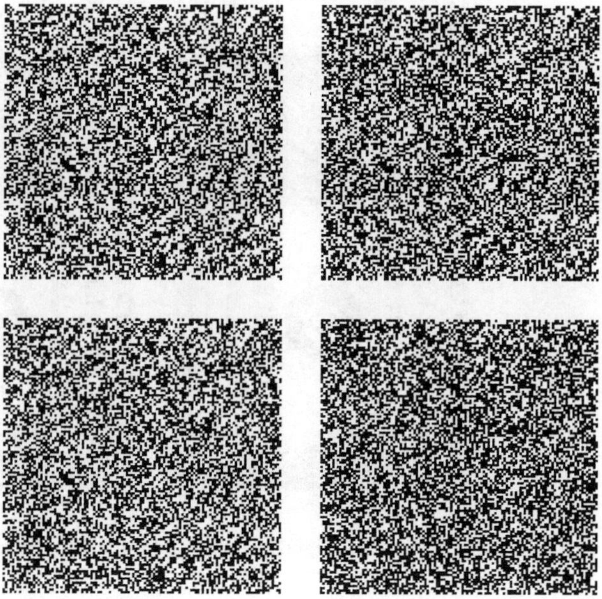

Fig. 6.13 Recognition of noisy stereo information: A floating triangle can be seen both above and below, but with increasing difficulty

6.3.5 *Where's the Mistake?*

The ability to combine individual images can be used to create astonishing effects. For example, two identical looking images can be examined for errors or a banknote can be exposed as a counterfeit. As in Fig. 6.14, place a real banknote next to the banknote in question.

The search for the error is very tedious if you don't use the simple trick of stereo viewing: Fuse the two images in the same way as for the random dot stereograms. Now the errors clearly emerge as unstable points of ambiguity in the depth, and the falsification is exposed. A total of three errors can even be detected in Fig. 6.14.

If you understand how to use the stereo view correctly, this ability is sufficient to make your fellow human beings enthusiastic. This can even bring you to TV show appearances where people show their "magical" talent.

Fig. 6.14 On the left is the original, on the right is the fake: Where are the three errors? The solution is very simple with the stereo view

6.3.6 *The Rivalry of Structures*

The purposeful striving of our perception to unite the double images is, however, only one side of the coin. If we extend our observations to more structured images, we very soon recognize a second, opposing and equally strong, principle of nature. To do this, look at the next stereo image pair in Fig. 6.15.

The result of the experiment with Fig. 6.15 is a little surprise: There is no mixture of the two bars. The two beams are in strong rivalry to each other. Therefore no stable compromise can be found. Our perception thus resorts to a trick that we have already investigated in a previous journey: the temporal change of alternative structures!

The horizontal bar alternately interrupts and hides the vertical bar, and vice versa. This is mainly a competition between the borders involved.

Of course, this competition between structures is again influenced by the Gestalt laws. It was found that a retinal image with a well-arranged structure dominates one with a less well-arranged structure. In other words, the more even image is suppressed by the more complicated one. In addition to other design laws, the length of the boundary line of the structure can be taken as a yardstick for the strength of the structure—this is quite clearly evident from Fig. 6.16.

Fig. 6.15 The rivalry of structures: Do you see a vertical or a horizontal bar? Do you see them both together, or always alternately?

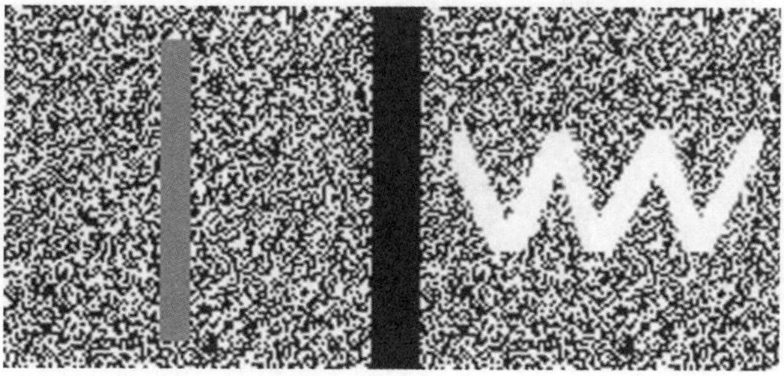

Fig. 6.16 Influence of design laws on the competition between structures: The right eye dominates over the left eye by the strong structure of the white

6.3.7 The Rivalry of Colors

Another important factor influencing the outcome of the competition between rival structures is color. It can spark new rivalries, but it can also allow for new compromises. Its role is very strongly connected with its supporting structure. Differently colored, rival and unstructured surfaces do not unite and the two colors do not mix—rather there is a rivalry between the two colors involved. This manifests itself in an apparent iridescence of the depth image. If this effect does not occur, one of your eyes strongly dominates the other in its vision.

If, however, the colors are supported by a strongly structured structure, the competition of the colors fades away. The double picture unites and the colors mix. You can check these two structure-dependent color competitions in Fig. 6.17.

In this stereo image, you can see that the competition between red and blue can produce very different results. The background, which is strongly structured, appears in the mixed color violet after stereoscopic union.

The less structured semicircles, on the other hand, seem to "shimmer": there is a competition between the two colors. In irregular alternation, the two colors appear alternately. If this iridescence does not occur, one eye strongly dominates the other. If the upper semicircle appears to you rather red than the lower one over a longer period of time, your left eye dominates. Experienced the other way round, the right eye dominates.

6.4 On the Main Road

Our coach through the fantastic wonder world of perception is now about to turn into a main road. After the strange and unfamiliar forms of the random dot images, we have already arrived at familiar patterns such as banknotes and seeing in color. So we are well on our way back to our familiar surroundings.

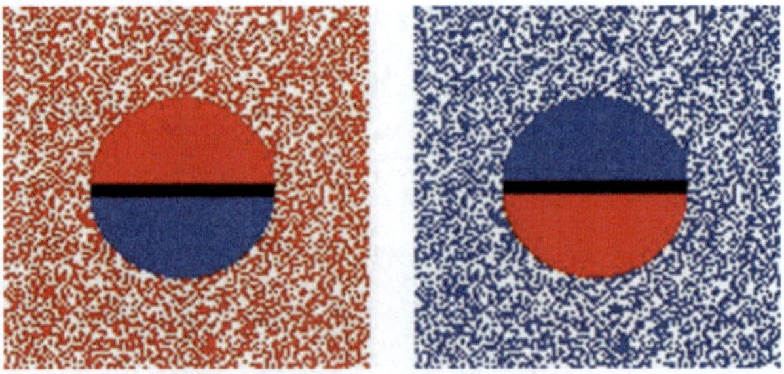

Fig. 6.17 The competition of colors

6.4.1 Stereoscopic Photography

In the following we will look at stereoscopic images of our familiar environment. These images are normally taken with an (expensive) stereo camera. This special camera consists of two combined cameras with a distance of 6½ cm—this corresponds to the average human eye distance. The shutter of the two cameras is coupled so that moving objects such as waves, flames, or moving animals can be captured simultaneously.

In order to understand the principle of stereo photography, however, it is quite sufficient to take two pictures of one and the same motif with a very ordinary "mono" camera. The positions of the two images should be shifted by at least the interpupillary distance. However, you should make sure that you only take pictures of still objects. Our experience with the stereo view gained with the random images does not make it difficult to fuse the individual images into a depth impression.

To do this, look at pair of pictures in Fig. 6.18 with the stereo view. After it has "clicked" in your perception, you see the mountain hut in a completely new "light": the whole scenery gets depth. A lot of small details are only now really noticeable. You can go for a walk in the depth of the image. Also, look through the window at the distant mountain panorama.

When you see such a picture for the first time, you wonder why stereo photography lives such a wallflower existence in our time.

Fig. 6.18 A stereo photograph. Enjoy the depth impression of this natural stereogram with the stereo view. Also take a look through the window of the ruin

In addition to the difficulties in learning the stereo view, there is a second important reason that prevents the stereo images from spreading further: After all, stereo vision is much more demanding than stereo hearing.

So far we have only got to know a small part of the spectrum of the possibilities of our perception to obtain depth information. This will change in the following sections.

6.4.2 *The Hollow Mask*

How complicated stereo vision can be, and with how many special rules, can be seen by looking at Fig. 6.19.

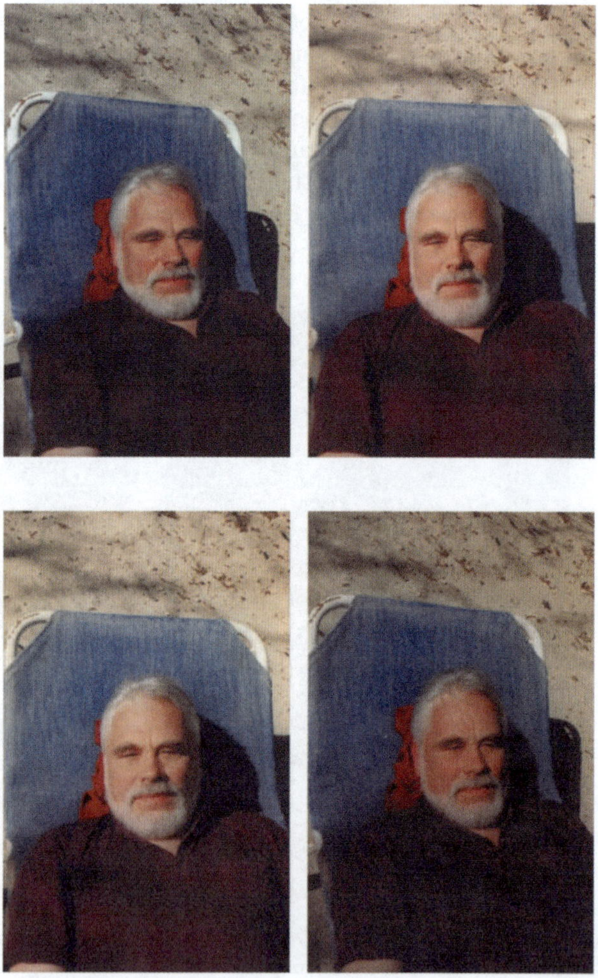

Fig. 6.19 A completely normal stereo image pair of a head (above) and a hollow mask (below): Despite the reverse arrangement in space, we also perceive a "forward curved" head at the bottom

Look at the two stereo images of the head. The individual images of the upper row combine very easily with the stereo view to create a depth impression. The face looks at us, and clearly the nose arched forward is to be recognized.

Even if you look at the pair of images below, you will get the impression of a face that is curved forward. However, a strange feeling remains, as if something is wrong with the picture. In fact, the image pair is a hollow mask! In reality, the face is spatially curved inward. This was achieved in this picture by swapping the two frames. Now the left eye is supplied with the image that was actually intended for the right eye, and vice versa.

According to the simple laws of depth perception that we have learned so far, we would have to clearly recognize a hollow mask from the transverse disparity of the individual images. Why isn't that the case?

The answer is: Our brain travels on several tracks during depth perception. Over millions of years, it has developed a number of "success strategies" to identify and assess depth. *Experience* is very important in this respect. We have learned that a face can never be curved inward. Therefore, the experience in this picture functions as a kind of "doorman" for the incoming depth information. The information "arched inward" is simply not passed on to our brain, the learned experience "arched outward" overrides the physically gained impression of depth. The decision dictated by experience is finally "perceived."

One way to circumvent this censorship is to alienate the familiar perception of "face." You can easily do this by turning your face upside down: Simply turn the book upside down and look at the hollow mask again. If you allow yourself enough time, you may be able to play a trick on your brain and recognize the hollow mask.

If you do not succeed, you should not doubt your ability to perceive. Rather, it is a "safety filter" of your brain, which should protect you from unrealistic perceptions such as hallucinations.

This filter must be in some way closely connected to areas of consciousness. This is shown by studies on persons suffering from acute schizophrenia or under the influence of drugs such as marijuana or LSD. In these cases, people can easily recognize the deep hollow world.

You can also reproduce the impressive experiment with a hollow mask in real life. Look at a carnival mask from the inside. If you close one eye, you will quickly see a normal, outwardly curved mask. The whole thing is particularly impressive if you now walk around the mask. This gives the impression that the mask moves with you and looks after you.

The influence of experience on the perception of depth is of course not limited to faces. Landscapes, plants, animals, fellow human beings, and other familiar things of our everyday life are also "protected in depth" by our experience.

What exactly is this experience of depth perception all about? Our brain knows from its experience exactly what some things must look like; it has, so to speak, stored "prejudices." In addition to sensory methods such as scanning and hearing, it also makes use of the known laws of sight—the brain, for example, compares sizes with known objects, evaluates light conditions such as brightness distributions or shadow casts, and observes overlaps of the body further back by the body further forward.

6.4.3 Recognizing Depth with One Eye

Surprisingly, people can do very well without three-dimensional vision. Try, for example, to start your day with just one eye! It is best to carry out this test immediately after the double image test when you wake up—this way your environment will be surprised at least only once about your recently strange methods of waking up! Even if you keep one eye closed, you will succeed in getting up, dressing, fetching newspapers, and reading them quite easily. Only with a few small things, like pouring coffee into a coffee cup a little further away, will you have a little more difficulty than usual.

Examples of one-eyed people doing very well in their environment can be found everywhere, even in the area of peak performance. The one-eyed German soccer player Wilfried Hannes (58 goals and 261 German Bundesliga soccer matches for Borussia Mönchengladbach) and Wiley Post, who made the first solo flight around the world, became famous.

How are such outstanding achievements possible? The reason for this is that although binocularity is the cleanest way of perceiving three-dimensional patterns, it is not the only way for our perceptual apparatus to recognize them. On the contrary, a variety of other strategies have developed that enable us to recognize our position in the surroundings. The reason for the diversity of these existing methods is certainly to be found in the millions of years of evolution of the visual system. The complicated procedure of depth measurement by means of transverse disparation required a very long development time. At the same time, some much simpler strategies have developed.

The use of additional vision strategies has several advantages: First, they step in when the second eye fails, for example due to injury. They are also needed when the depth resolution of stereoscopic vision, which is limited by the low interpupillary distance, is about 1 km. At these distances, these other strategies of depth detection are used without exception.

In addition, of course, they are used for stereo vision to check sensory impressions—we could already see this when looking at the hollow head. Only when we come to the same result with all visual methods, a clear visual experience is possible without having the unconscious feeling: "Something is wrong here"!

In the following we present a small selection of the most important concepts of stereo vision for a single eye. These are motion perception, the recognition of overlaps and transparency, the estimation of the size of known objects, the recognition of the shadow cast, and brightness perception.

6.5 Other Methods of Depth Perception

6.5.1 The Perception of Depth Through Movement

Move your head back and forth once and register how the objects in front of you seem to move. For example, the table in the foreground moves much faster than the wall behind it. And the foreground shifts more than the background. You can also put it another way: objects at different distances move with different parallaxes.

This apparent movement of spatial objects was also mainly responsible for your one-eyed orientation when you got up in the morning. Naturally this principle also works at faster movements: So you have surely already admired the pulling of clouds from the window of a fast-moving train. The lower cloud layers seem to race under the higher ones. Conversely, we can of course also estimate distances from moving objects when we and our eyes stand still.

From the speed of cars driving past us, for example, we can very easily deduce their distance. The closer they are to us, the faster they move within our field of vision. Thus, a cyclist can seemingly overtake a car with ease if he only rides on a road that is much closer to us than that of the car.

6.5.2 Depth Perception Through Detection of Overlaps

This simple principle is given by logic: Objects in the foreground cover those located in the background. Our visual apparatus therefore only checks which object covers which other and receives the respective depth information from this. The covered objects are then completed in our perception according to the experience and the Gestalt laws. An example can be seen in Fig. 6.20: In the left half of the picture you will probably first recognize an arbitrary arrangement of violet patches.

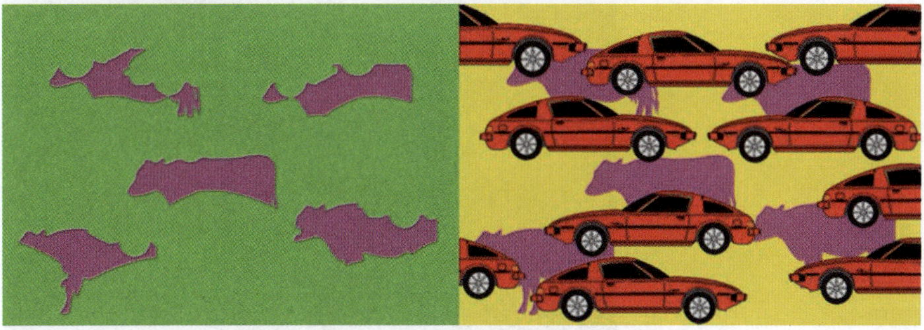

Fig. 6.20 Depth assignment based on overlaps: In the left half of the picture, only the uncovered parts of the picture are shown and appear as randomly arranged patches. With the help of the additional spatial information, our perception completes the covered objects to cows

Only with the additional information from the right half of the picture about the overlapping objects the patches are completed to clear structures.

6.5.3 *Depth Perception Through the Recognition of Transparency*

We already got to know the property of the recognition of transparency during the third journey. Therefore, we only want to deal here with the effect in terms of depth perception. Logically, the transparent object must always appear in front of the object being compared, which shimmers through the transparent object. According to this principle, we can, for example, assign different depths to car windscreens or window panes.

An example of this can be seen in Fig. 6.21, taken in Boston.

Fig. 6.21 The depth perception criterion of transparency: The window that appears "transparent" is in the nearer foreground

6.5.4 Depth Perception Through Size Comparison

Another simple tool for estimating depth is the size comparison. For this we use everyday experience about the size of known objects: The smaller the object appears in the field of view compared to other known objects, the farther back it seems to be, and vice versa. We got to know this phenomenon of size constancy already in our second journey. If we rely only on our sense of depth by comparing the size of things, we get into trouble when the conditions are no longer right or when objects appear whose size we do not know.

Underwater, for example, distances can be easily misjudged, because water has a different refractive power than air. Therefore, our values of experience are no longer correct; depth perception due to the size of known objects is significantly changed. To prove it, try the following: immerse your hand in an aquarium or in a hand basin. It now looks a lot bigger than before. Because of the refractive power of water, the hand appears to be closer than it actually is—as can be seen in Fig. 6.22. The opposite effect occurs when you completely submerge your eyes in water, as in a swimming pool. Your hand now looks a lot smaller than before—even when there are no piranhas in the pool!

For the same reason you become an apparent giant when you look at yourself underwater down to your feet. This is why other bathers or the edge of the pool underwater appear farther away than they actually are.

But even very simple deviations from everyday experience can cause our sense of depth to sway considerably with the help of size comparisons: When looking at a puppet show, for example, you can get a real scare as soon as you see a hand of the puppeteer on stage. This is due to the fact that our depth perception suddenly collapses with the help of size comparison.

Other examples of such "aha" experiences are pictures of miniaturized cities in theme parks or the backdrops of animation film studios in which a person of the right size can be seen in the background. A major cause of the great attention that

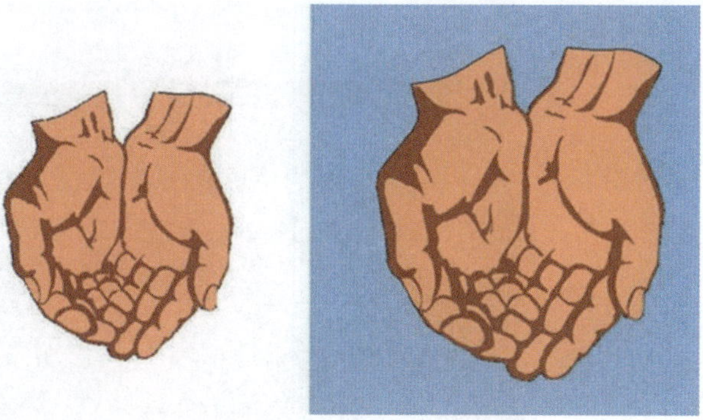

Fig. 6.22 Due to the different refractive power of water compared to air, our hand appears larger in water than in air. That's why it seems closer underwater than it really is

very short or very tall people attract is precisely this collapse of our deep-sightedness through size comparison associated with their occurrence.

The complete failure of the depth determination for some holiday pictures has its cause in the absence of known everyday experiences about the size. Without known sizes in the field of view, of course, you can't compare sizes. An example of this can be seen in Fig. 6.23.

This problem of assigning the right depth to unknown motifs is often solved by including people in holiday pictures: "Stand in the margin, please, for size comparison."

But this does not explain by far the recording technique that is mainly used by tourists, who often hide the attraction in their photos taken by the persons in the foreground.

A strategy for depth detection with one eye that is as complicated as it is effective is the observation of the varying influence of light on objects. Light can create differences in brightness and shadows, the meaning of which we will examine in more detail below.

Fig. 6.23 Upper part: Large rock formation or sandbox? Holiday photo without size comparison. That is why it is difficult for us to assign the correct depth and size to the picture. Below: Holiday photo with size comparison!

6.5.5 *Depth Perception Through the Interpretation of Shadow Casting*

All spatial structures normally cast a shadow—except for the (in our case rather rare) moments in which the sun shines exactly from above. It is a known fact of observation that the depth effect is most intense with strong shading. This is of course hardly the case with vertical solar radiation. The best light for detecting spatial depth is therefore in the early morning and late evening hours. Then the shades are pronounced and the perceptions become vivid. For this, look at the two fundamentally different images of the beach scene in Fig. 6.24. Both were taken from almost the same position. However, the left picture was taken at noon and the right in the evening. The differences are enormous: The left picture has very little depth and seems boring, even lifeless. The picture taken on the right in the evening has much more atmosphere. The interplay between shadow and light makes the shore formations much more vivid.

The grazing evening light, in connection with shadow-casting, is thus much better suited for depth perception than the midday light. Besides the better shading, the evening light seems "softer." The reason for this is that it is no longer as intense and is accompanied by much less disturbing scattered light than during the day.

The different effects of midday and evening light can be observed very well on lunar images, as in Fig. 6.25—and summarized in a single image.

Here the sunlight falls from the right onto the lunar surface. In the zone at the right edge of the image, the light falls vertically. One can clearly see that this area has hardly any contours and is almost uniformly brightly illuminated. The contours of the existing craters are virtually invisible. This is the exact equivalent of the light conditions on Earth at noon.

Now look at the light conditions at the left edge of the illuminated moon: Due to the grazing incidence of light on the lunar surface, the craters in this area are now very clearly visible. The surface gets spatial depth and appears very varied and interesting. This again corresponds exactly to conditions on Earth in the evening.

Fig. 6.24 A beach scene, daytime and evening

Fig. 6.25 The effects of the different position of the sun on depth recognition: At vertical solar incidence in the brightest lunar zones, contours and craters are hardly visible. On the other hand, these stand out very strongly with grazing incidence of light at the edge of the illuminated moon part. The lunar surface appears plastic in these areas

The shadow cast in combination with the important fact of experience that the sunlight comes from above helps us to recognize the spatial depth—as can be clearly seen in the upper part of Fig. 6.26.

In both the upper and lower photographs, you will surely recognize a winding stone pattern protruding from the wall. But there is a big difference between the two patterns: At the top the pattern appears as a continuous band, at the bottom it is interrupted again and again. In reality, the two images are completely identical, only turned upside down. Turn the book upside down to check.

Explanation: Experience tells us that the light of the sun always comes from above. In the two pictures, the different direction of the shadow cast initially gives rise to different spatial assessments. However, our perception system knows that only protruding stone parts provide a shadow cast.

According to the assumption of our perceptual system that there is vertical solar radiation, this shadow must lie on the side facing away from the sun, i.e., below the protruding pattern. This results in the respective estimation of the image depth for both images.

The same explanation is the reason for the depth deception in the lower part of Fig. 6.26, taken at a platform of the Prague subway. Surely you can see on the left a row of round semicircles arched outward. With one difference: In the lower row the semicircles are curved inward. These spatial conditions change drastically as soon as the image is turned upside down as shown in Fig. 6.26, below right: What was

Fig. 6.26 Certainly you will see a continuous stone pattern in the upper row in the upper picture. On the other hand, the pattern underneath seems to be interrupted again and again. In reality, these are identical images turned upside down. A similar phenomenon can be seen when you look at the two images below. In the original picture on the left from a subway station in Prague, most of the semicircles appear curved outward. This impression changes completely when the image is rotated by 180° (see image on the right)! Turn the book upside down to check

previously curved inward now appears curved outward and vice versa. Again—also within a sunlight-free underground shaft—the virtual sun/illumination is assumed to come from above. The rest is done by our perception.

The reason for the different depth perception lies in a fundamental everyday experience about the position of the sun: Our perception automatically assumes that the sunlight falls from above and not from below.

6.5.6 Depth Perception Through Brightness Contrast Detection

Brightness contrast also plays a crucial role in assigning depth. Thus, objects rich in contrast seem to be closer to the viewer than objects with less contrast. This phenomenon again originates from our everyday experience: scattered light and clouded air make distant objects appear increasingly blurred and out of focus. You can see this effect in Fig. 6.27: the cows shown seem to lie at different depths. Surely the high-contrast cow at the left appears to be in the foreground and the blurred right cow seems the farthest in the background. Due to the principle of size constancy, the right cow also appears to have the largest perspective. In reality, all the cows are exactly the same size.

The experience-based conversion of brightness contrast into depth perception is also the cause of misinterpretations of distances from distant landscapes. For example, the Alps—seen from the foothills of the Alps—appear to be within reach on some days and very far away on others, although their distance is of course always constant.

The simple explanation: With a clear view, the mountains appear only slightly blurred and very rich in contrast. Therefore, we estimate their distance to be very close. In contrast, in cloudy, foggy weather with poor visibility the mountains are only blurred; they are therefore perceived to farther away than they really are.

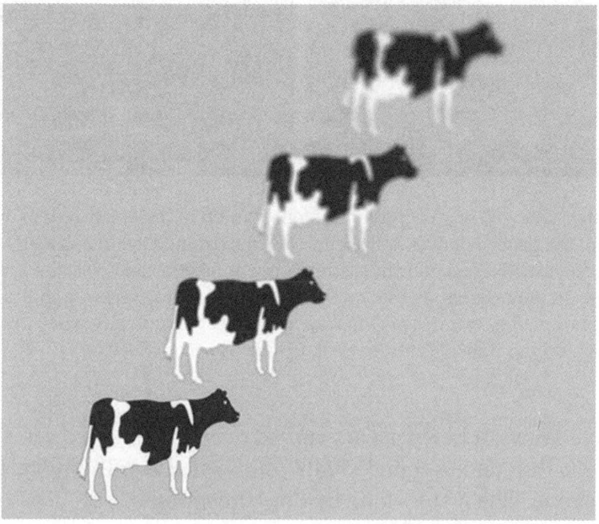

Fig. 6.27 Depth sensation by comparing the brightness contrast: Which of the four cows of the same size appears closest? The high-contrast left cow appears in the foreground, the blurred right cow in the background. Due to the principle of size constancy, the right cow appears to be the largest in perspective

6.5.7 *An Unrealized Method for Depth Perception*

Why is the recognition of spatial depth not actually done by measuring the curvature of the lens? After all, the lens of our eye does cause the focusing—the *accommodation*—on any distant object in the same way as a camera lens. In the remote distance setting, the lens is flat and only slightly curved—this is achieved by tightening the external ciliary muscles; in the near setting, however, this muscle is relaxed accordingly and as a result the lens surface is strongly curved.

It would therefore be logical that sensors exist in the ciliary muscles that would test the tightening of the muscles for the purpose of depth measurement. However, nature has almost completely done without it; only in two extreme cases do we get a sensory message about the state of the accommodation of the eyes: in the case of a very strong distant adjustment, a feeling of relaxation arises; in the case of a very strong close-up adjustment, however, one soon feels strained.

For the entire intermediate range we do not receive any sensory information about the condition of the lens curvature. Nature deliberately refrains from extracting depth information from the accommodation of the eyes. The reason for this is probably that the extreme distance adjustment of the lens is already realized with the eye from 5 m.

Already from 2 m on, only minimal influences of different lens curvatures on the image sharpness are recognizable—therefore, a measurement of the accommodation would only make sense in a range up to 2 m anyway.

With our knowledge of the additional methods of depth perception gained in the last pages, it also becomes clear why we found the depth resolution values given in Table 6.1 to be too low. These values refer exclusively to the resolution due to transversal disparity. Our depth resolution in everyday life, on the other hand, consists of the sum of all the depth perception methods presented, which leads to significantly better values than those given in the table, especially for long distances.

6.6 Why Do People Have Two Eyes?

After many sensory errors and confusion, our travel group has now actually returned to its starting point. Just like our itinerary, the way our depth perception works is also quite confusing. As you can see, just like on a journey, there are many ways to reach your desired destination. But it is only the sum of all these paths and visual strategies that enables us to enjoy perfect deep vision.

Indeed, it seems that we have not come much closer to answering our initial question about the number of eyes needed. There was a plethora of answers, but also a plethora of new questions. We have seen that the miracle of depth perception consists of a whole network of interwoven strategies and reaction loops that can sometimes contradict each other or end in dead ends, but which usually produce very clear solutions.

The assumption suggests itself that it is precisely this confusion of methods that produces the outstanding qualities of our depth perception. For what system would be more suitable for better representing our often confusing and contradictory three-dimensional environment than this almost equally confusing, self-organized system of rules governing our perception?

The second eye is only at first sight a kind of "luxury" of nature. Just like the second look, the second eye is well worth it. We noticed this several times during our journey: on the one hand as a replacement for the loss of one eye, of course; on the other hand to guarantee the wonderful visual performance with the help of the transversal disparity of the spatial objects.

A third eye would be nice, but given the need for proper wiring within the visual system—just think of the 126 million visual receptors—a very large amount of evolutionary effort would be required, and would not result in an appropriate new quality.

So let us rather continue to accept that someone can sneak up from behind without us noticing him, and that we cannot see around the corner. Because that's where life's little surprises come from. It seems that nature has found an optimum between functionality and resource consumption with the number two.

But maybe the real reason why we have two eyes is simply so that we can wink with one eye every now and then!

6.7 In Venice Beach

Whole groups of people crowd like wild before colored pictures. Screams of joy and a thirst for adventure dominate the somewhat strange scenery. It is a "first time mood" among people, like before the first attempt to swim without inflatable floats, before the first attempt to ride a bike without training wheels, or before the first parachute jump. You don't know exactly what to expect or when something will happen.

There! At the very front a place becomes free, and suddenly you stand before a picture on which is written in big letters: "Relax!"

"No panic," would be more appropriate in view of the now wildly gesticulating and discussing crowd. Next to you and around you, it's about how to describe the feeling that comes "the first time"—a feeling that you would like to have now as well.

I heard a loud "plop," says the fat, sweaty gentleman behind you.

You look at the picture again and wonder what "Plop" is supposed to do here. All you ever see are those colored circles with no sense, nothing more. Is that all you've got? And what's so special about that? You remember the story about the emperor and his new clothes, which didn't even exist, but there can't be emperors all around you. Who is this story from again? And how did it go, exactly? Suddenly the word "dolphin" appears in your subconscious. What does that mean, is there any connection with the emperor just now? Just now? Somehow time stands still!

And suddenly a wall of glass is built up in front of you, which comes closer. Behind it a deep abyss opens. You feel like in a dream, only with the difference that you can watch yourself here. Just don't think now! The glass pane is approaching alarmingly fast, and there it happens: The pane shatters into a thousand pieces, you will never forget the first moment afterward! Somehow you feel weightless. Apart from that you are the same—but suddenly everything happens inside you like underwater or in slow motion. In all peace of mind you can now go for a walk in the depths, even to the bottom.

A dolphin suddenly appears in front of you. I don't know why dolphins so often sit in these 3D pictures—maybe because they are so intelligent! After all, you know now: you are in a 3D image and not in some dream!

I wonder what time it is right now. Never mind! To find out, you would have to leave your new world—but why? Just at this moment you feel a bump and a man next to you asks you: "Dolphin? Why do you always yell *dolphins*?" All of a sudden the spook is over—there they are again, all the people around you, the sun is shining in the meantime. Hard to believe: You are already standing here for half an hour.

"When you hit the glass, it's gonna go, 'Zack!'", you hear yourself say to the seemingly perplexed man next to you.

This story took place in 1993 in Venice Beach and is about one of the most famous flagships of perceptual psychology: the *autostereograms*. They experienced a tremendous boom around the turn of the millennium. Never before had harmless, originally scientific images succeeded in triggering such enthusiasm around the globe.

6.8 A Time Travel Through the Technique of Stereo Vision

In front of our travel group stands a square, metal thing without wheels, which at first glance does not have much in common with the fantastic adventures promised in this book. After all, it has a gangway, via which we get into the interior of the machine. Like in a cinema, some armchairs are placed in the middle of the room. The bus driver sits next to us and seems somehow grumpy. He would rather be at the wheel himself, but the itinerary now provides for something else. Above us a lettering appears: "Welcome to the fantastic time-travel through the technology of stereo vision." Aha, we are in a time machine!

And already the whole room begins to buzz, and a digital counter shows us the years: 2005, 1995, 1950...

"And now we are in 1838," says the bus driver, bored. After all, this is the nineteenth time he's been on this tour.

6.8.1 The Mirror Stereoscope

This is the year in which our time journey into the history of stereo vision is to begin. Although Euclid, Leonardo da Vinci, and René Descartes have already dealt with the problem of the representation of the third dimension, the actual time calculation of the understanding of 3D perception begins in 1838.

The optical workshop of the British inventor Charles Wheatstone is just appearing in front of our seats. He was the first to find a way to view images stereoscopically. He achieved this with the help of the *mirror stereoscope* he developed, the structure of which is shown in Fig. 6.28.

As the name suggests, this viewing device consists of a mirror apparatus. With the help of these mirrors, it is possible to feed the viewer's eyes with two slightly different images on a separate beam path. This makes it very easy to combine the two images into a three-dimensional impression. The only prerequisite is that the mirrors are precisely adjusted and that the images are arranged at the correct angle and distance from each other. The images considered were artificially created drawings (as in Fig. 6.29), which differed from each other only in a horizontal deviation—the transversal disparity.

Without stereoscopic observation, Fig. 6.29 can be seen in at least four different perceptual alternatives: One can see a vertical reel of yarn, a bow tie for the theater evening, and two perspective views—a tunnel open to the back and a truncated pyramid projecting upward with a square base. When viewed stereoscopically with the Wheatstone mirror stereoscope, however, the ambivalence suddenly dissolves and the truncated pyramid protrudes far from the sheet level.

The mirror stereoscope facilitates stereo vision by the following trick. The two images are spatially far apart, but lie at the same depth. The beam paths between the image and the eyes are deflected by the mirror system. The length of the beam path is set exactly to the spatial distance specified by the transversal disparity by a clever adjustment of the mirrors. Thus the two pictures appear lying at the right place in the right depth.

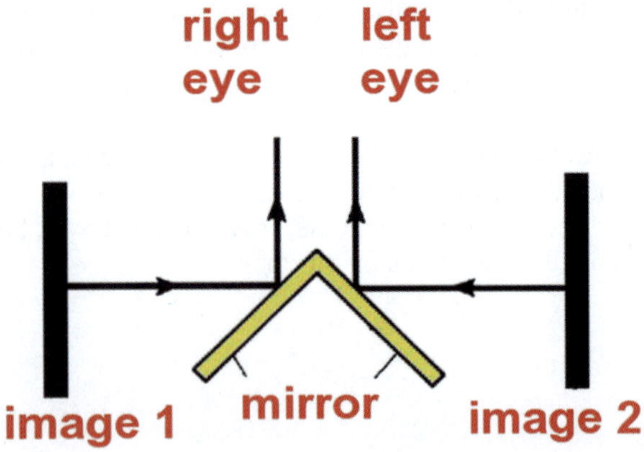

Fig. 6.28 Wheatstone's mirror stereoscope from 1838

Fig. 6.29 Example of an artificially generated image pair that can be viewed spatially with a mirror stereoscope: Without a stereoscope, the image can be seen in at least four different perception alternatives: Thus one can observe a vertically standing reel of yarn, a bow tie for the theatre evening, and two spatial perspectives—namely a tunnel open to the rear and a pyramid stump projecting upward with a square base. When viewed stereoscopically, however, the ambivalence suddenly dissolves and the truncated pyramid protrudes far from the paper

6.8.2 The Lens Stereoscope

Shortly after the invention of the mirror stereoscope, David Brewster achieved a great improvement. He replaced the mirrors with a special lens system and developed the *lens stereoscope*. This made the device much easier to handle and accessible for commercial use.

Stereoscopes manipulate the visual beam so cleverly that we don't have to learn any new visual technique to recognize stereoscopic images. Either the beam path or the focus is adjusted accordingly. This leads to a completely normal visual experience.

In the lens stereoscope two lenses are installed in front of the two eyes with the exact refractive power to widen the visual rays by the correct angle and only cross far behind the image plane. The structure of such a lens stereoscope can be seen in Fig. 6.30. The stereo images are arranged next to each other in one image plane. In addition to being easier to handle, the lens stereoscope has the additional advantage that its stereo images can be printed in books such as this. Once one has learned this technique to decouple the focusing and viewing angles, these images can also be seen with the naked eye.

6.8.3 Visual Techniques With and Without Stereoscope

The stereoscopes support our eyes in the effort to achieve the stereo effect. The difficulty of stereo vision with the naked eye is to bypass a mechanism of vision learned from childhood. This mechanism is useful when detecting all "normal" objects: the visual acuity of the individual eyes is coupled with the adjustment of the two visual axes.

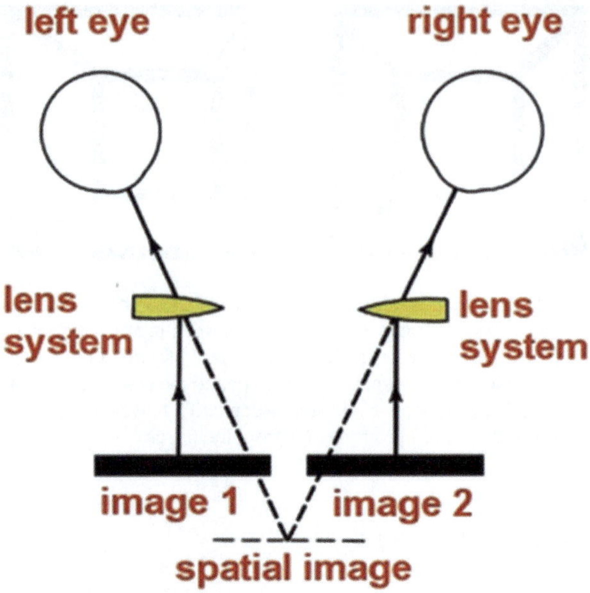

Fig. 6.30 The principle of the lens stereoscope

We first fix any point in space by adjusting the visual directions of the two eyes in such a way that the visual axes intersect at this point—this is the convergence of the visual axes, which we have already discussed. The lenses of our two eyes automatically focus on this point of intersection, which is called *accommodation*. For normal spatial vision the following applies: Accommodation and convergence are coupled.

Our strategy for *stereo vision*, on the other hand, is completely different: The individual eyes should continue to focus on the image plane. But the gaze should not be directed at the sheet. Rather, the axes of vision should meet outside the plane of the leaf, whereby the distance to the leaf plane determines the perceived depth. This point of intersection of the visual axes can be located both in front of and behind the plane of the leaf. This results in two different possibilities of stereo vision: *staring* and *squinting*.

More detailed information can be found in Fig. 6.31. When staring, the view is directed behind the image plane. This possibility (left partial image) has its limits in the interpupillary distance: Only pixels at a distance of less than 6.5 cm can be combined stereoscopically—but this is possible with most images. The view is very pleasant, gives a relaxed feeling, and can be maintained for a very long time.

Squinting, on the other hand, requires some effort and should not be exercised for too long. The depth information is exactly reversed, because the visual rays meet in front of the image plane. The advantage of squinting is that even image pairs with an arbitrarily large distance of more than the interpupillary distance can be aligned.

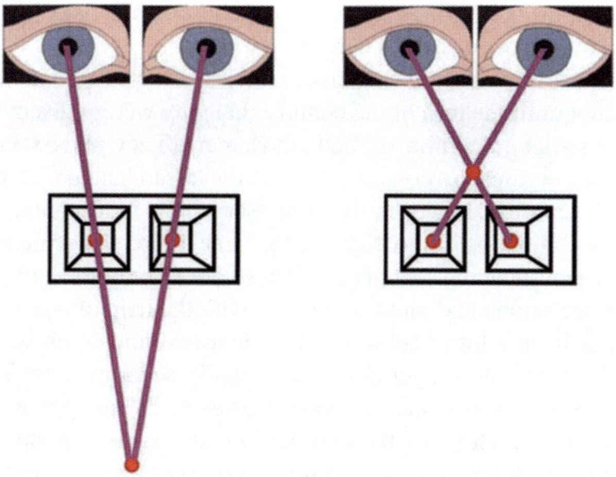

Fig. 6.31 The two possibilities to see stereo images: *staring* and *squinting*. When staring, the gaze is directed behind the image plane; when squinting, the visual rays meet in front of the image plane

During this long explanation, the display of our time machine changed to "1839." In this year, the French painter Louis Daguerre succeeded in inventing *photography*. This had far-reaching effects on the further development of stereo images. It was now possible to take and see natural stereo photographs in addition to the artificially generated images. Soon there were special stereo cameras, which consisted of two cameras connected at eye distance. This new technology enabled an even faster spread of stereoscopic images.

The 1851 World Exhibition in London increased the public's interest in stereo vision, and in 1856 stereo images and stereo viewers were available on the market for the first time. Finally, the London Stereoscopic Society was founded in 1893, and still exists today.

6.8.4 The Wallpaper Effect

In the meantime, our journey through time arrives in the year 1844. The discovery of another effect that was important for stereo vision, and the development of autostereograms, is dated to this year. Again, it was David Brewster who succeeded. One day he noticed something amazing in a Victorian wallpaper printed with a constantly repeating motif: The wallpaper turned into another depth as soon as he looked at it cross-eyed! Therefore, this effect is called wallpaper effect. It is a direct precursor of today's autostereograms.

You can follow this depth transformation on a wallpaper and on many other everyday things. The only requirement is that you have to look at a periodic pattern. As soon as you have mastered the staring technique or the squinting technique, you can see roof tiles, bridge railings, car ceilings, chess boards, skyscraper windows, bathroom tiles, or fences in various depths.

Shortly after its discovery, the wallpaper effect was forgotten—as were the stereo images. It was not until the turn of the century that they were rediscovered in aristocratic circles as social entertainment and amusement. They played a role comparable to ambiguous figures or comics. Due to the disadvantage of the visual aid necessary for observation, however, they only served the amusement of small affluent social circles. Finally, stereo pictures became more widespread through the development of new viewing techniques. These are mainly the red-green or blue-green anaglyph technique and picture-viewing with the help of a *polarization filter*.

Both methods have a lot in common. The decisive innovation was that the two individual images were no longer depicted spatially separated, but on top of each other in the same place. For stereo vision, however, the images must be viewed separately by each eye. How are the two individual images kept apart? Both techniques use different filters for this purpose. If you put them in front of your eyes, only the picture that fits in each case is let through to the corresponding eye, and the other component is faded out. The main difference between the two techniques lies in the type of filter used.

6.8.5 The Red-Green Anaglyph Technique

In the anaglyph technique, the wavelength of the light, i.e., the color, is used as an aid to keep the two individual images apart. The creation of a red-green image is done as follows: The recorded stereo images are colored in different colors—one image in green and the other in red. These two monochrome images are superimposed with the correct transversal disparity. With the naked eye only an uninteresting double image can be seen, which looks like a misprint. If, on the other hand, the picture is viewed with red-green glasses, the stereo impression is created in a very simple way. One eye is held by a red filter and the other by a green filter. If you do not have red-green glasses in your household, you can easily make a pair yourself from colored tissue paper. Use it to view the red-green images in Fig. 6.32.

These fantastic pictures were taken by Franz-Josef Heimes with a special camera mounted on his aircraft, which he repeatedly triggers depending on the flight altitude and flight speed, the focal length, and the image format. Thus the houses of the German city of Olpe and especially the quarry in Fig. 6.32 below gain a completely new spatial depth.

The red-green pictures became known mainly through the cinema. This new invention of moving pictures was soon associated with stereo vision. A whole series of stereo films was created, which impressed with their very good depth effect. Unfortunately, most of them also impressed by unspeakably bad scripts—which was probably the reason why this wave of interest in stereo viewing faded again. Another crucial disadvantage of the red-green technique is that the depth images are not colorful.

Fig. 6.32 Two fantastic red-green pictures. Above: Aerial view of the city of Olpe, Germany. Below: Stentenberg quarry. (Courtesy of Franz-Josef Heimes)

6.8.6 The Polarization Filter Technology

This disadvantage does not apply to the viewing technique with polarization filters. To separate the two individual images, a special characteristic of light, which was already considered in the fifth journey, is used here: it can be polarized. This is a kind of secret code of light, which—in contrast to the bees—we cannot decipher with our eyes. Only with the help of polarization glasses we can see how and whether the light was coded. Only one of the two codes is let through by the glasses at a time.

We already got to know the physics of polarized light in the fifth journey. This was done by comparing it with a swinging rope. In this model, a polarization filter can now be imagined as a narrow slit that allows only those parts of the rope waves that oscillate exactly in the direction of the slit to pass through. The polarization filter for the second lens of the polarization glasses consists of a slit in a perpendicular orientation.

This technology is particularly suitable for use with projectors, for example in slide shows or films. The stereo projectors consist of two individual projectors with which the two stereo images are brought onto the screen. A differently oriented polarization filter is held in front of each of the two lenses of the stereo projector. These differently polarized single images can be fed back to our eyes separately with the help of polarization glasses. This process leads to stereo effects, with which color perception is now also possible without restrictions.

In no time at all, quickly and probably now irreversibly, 3D viewing has (finally!) become an integral part of everyday life. Driven by new animated movies from the big Hollywood studios, the first 3D blockbusters, such as "Ice Age 3" and "Avatar," were released in 2009 into the cinemas. The success of these films and the simple and inexpensive polarization filter technology have induced cinema operators to upgrade to 3D technology. As a result, 3D has almost become the standard for every new film and is about to make its way into our living rooms via 3D televisions.

However, this technique is completely unsuitable for printed products because, in contrast to polarized light projection of the two individual images, there is no possibility here to encode the individual images accordingly.

6.8.7 The Pulfrich Effect

There are two other methods to achieve a stereo effect in movies: The better-known and less elaborate of the two is based on the Pulfrich effect, which you got to know on the fifth journey. Incidentally, its discoverer, Carl Pulfrich, could not perceive his deception spatially himself—he belonged to the group of people who were blind in one eye (cf. Sect. 6.4.3)! As already mentioned, Pulfrich discovered that our brain perceives different brightness levels at different speeds: The brighter the light, the faster the visual stimulus is transmitted from the eyes to the visual center (Pulfrich 1922). Dark visual stimuli, on the other hand, are transported comparatively slowly. Like almost everything in nature, these different stimulus transmission times have a practical benefit. For example, very bright light, such as direct sunlight, poses a greater threat to the retina than dark light. A rapid protective reaction in the form of a contraction of the pupil is therefore required.

Such nerve lines with different speeds are not so rare in the perceptual system for a variety of reasons. We already saw in journey 5 that the color information "blue" is transmitted more slowly than the color information "red" or "green." Another example can be found in the human sense of touch: It has been shown that a gentle touch is transmitted to the center of perception more slowly than a tough blow!

You can observe the Pulfrich effect directly by darkening the perception of one eye. Simply hold a sunglass lens in front of one eye. Thus all visual stimuli of this eye arrive somewhat later in the brain than those of the "bright" eye. You can check this if you have a weight in front of you, such as an eraser, swinging back and forth from left to right on a string in one plane. The visual delay creates the clear impression of a three-dimensional elliptical movement.

This can be clearly seen in Fig. 6.33, where sunglasses are held in front of the right eye. The perceived orbital motion can be divided into two parts:

1. Pendulum movement from left to right: The point of intersection of the two temporally shifted rays of vision lies in front of the orbit level of the pendulum! That's why the pendulum seems to be moving toward the observer!

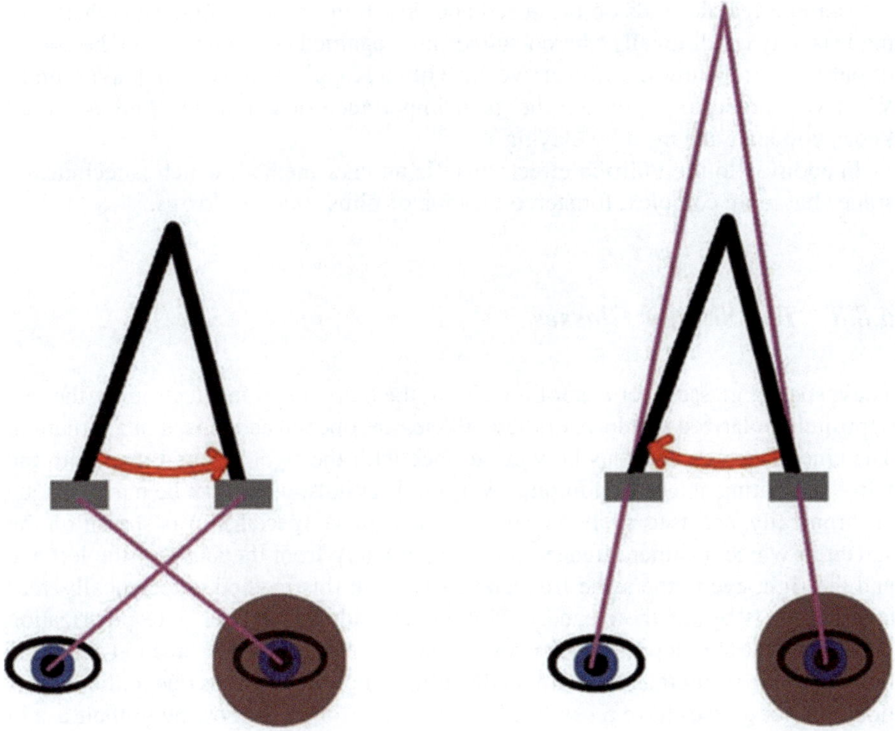

Fig. 6.33 The Pulfrich effect: Attach an eraser to a string and let it oscillate in front of you. Hold dark sunglasses in front of your right eye and observe the movement of the pendulum. This gives the impression that the pendulum no longer oscillates in a plane but describes an ellipse

2. Pendulum movement from right to left: The point of intersection of the two temporally shifted rays of vision lies behind the orbital plane of the pendulum! Therefore the pendulum seems to move away from the observer!

The Pulfrich effect can be used for a stereo sensation in the cinema. This of course requires a moving camera! The camera is moved at exactly the speed that compensates for the time delay caused by the Pulfrich effect. The visual information about an object thus arrives simultaneously in our eyes from two different positions (ideally from the normal eye distance). This makes a perfect stereo experience possible.

While driving slowly, record the view out of the window from the passenger seat with a video camera. If you watch the film with one-eyed, slightly darkened glasses, the Pulfrich stereo effect will kick in—of course, you will have to put the darkened lens on the correct eye depending on the direction of travel. If you have filmed a ride from left to right, you should have the dark glass in your right eye, and vice versa. You can also vary the film playback or driving speed to achieve the optimal stereo effect!

Stereo films based on the Pulfrich effect have been broadcast on the Discovery Channel, for instance, or in ads. However, this technique was not able to assert itself for the long term because it is associated with some major disadvantages: On the one hand, the camera must be in constant motion. On the other hand, the strength of the stereo effect depends on the speed and direction of movement. Thus this technique is only conditionally "stereo-suited" and regarded as a gimmick. In the course of our fantastic journey, however, we have often stopped at adventure playgrounds. We have learned to appreciate the great importance of gimmicks. And as we all know, you learn the most by playing.

In addition to the Pulfrich effect, there is another method, which is technically somewhat more complex, for stereo viewing of films: *shutter glasses*.

6.8.8 The Shutter Glasses

These special glasses work as follows: With the help of special electronics, the two oppositely polarized windows of these glasses are opened and closed in alternation. The cinema screen can thus be viewed once with the right eye and then with the left—alternating at a certain frequency. Liquid crystals, which can be made opaque electronically, are used as the closing mechanism. A special film is shown on the screen in which a camera image appears alternately from the angle of the left eye and the right eye at the same frequency. Although this method is technically very complex, it is by far the cleanest solution. The advantage over pure polarization glasses is that the views of both eyes can never be seen on the screen at the same time, with the result that there are no disturbing superimpositions due to the projection. Shutter glasses have recently experienced a strong boom owing to their use in most new 3D televisions.

6.8.9 The Random-Dot Stereograms

Already at the beginning of this journey we got to know the random-dot stereograms (see Sect. 6.3). This new form of stereoscopic image has brought a very important new insight to the psychology of perception: depth vision is possible without any perception of form—solely by means of transverse disparity!

Our time machine is currently travelling to the year 1959, when the discovery of computer-assisted random-dot stereograms is officially dated. Bela Julesz made this groundbreaking discovery in the Bell laboratory, which he perfected over the years. He and his coworkers, including a certain Christopher Tyler, were constantly creating new images.

6.8.10 The Autostereograms

This Christopher Tyler took the final decisive step in the development of stereo images for the time being. Together with programmer Maureen Clarke, he produced a few black-and-white images, called *autostereograms*, at the Smith-Kettlewell Eye Research Institute in San Francisco in 1979 (Tyler and Clarke 1990). These were no longer stereo image pairs consisting of two single image pairs, but only a single image! These autostereograms soon developed an unexpected momentum of their own. They were passed on, admired, and copied. They were improved, got colored structures, and increased rapidly over the computer networks and copying machines of the whole world.

In retrospect, the invention of autostereograms was an inevitable thing. The random point image pairs had the great disadvantage that they were difficult to perceive stereoscopically with the naked eye. In addition, it was an aesthetic disadvantage that these were two single images. These two disadvantages are fortunately linked in a special way. This will eventually lead us "by itself" to the autostereograms. This process can be seen in Fig. 6.34.

In order to recognize the random dot images with the naked eye, as in Fig. 6.34a, it is necessary to decouple the visual direction and the focus of the eyes. The extent of this decoupling is crucial in order to be able to see spatially. The deeper one has to look behind the image plane, the more difficult it is to decouple the eyes. In order to facilitate spatial vision, a simple strategy helps us: reduce the distance between the two images, as in Fig. 6.34b! The distance between the two images is best left out completely (as in Fig. 6.34c). The distance between the two images can easily be further reduced by narrowing the images themselves (Fig. 6.34d). In order to restore the picture to its original width, we recall the wallpaper effect: instead of a picture with only two repeating patterns, we complete the page with a whole series of stripes with periodically repeating patterns (Fig. 6.34e). Depth perception is greatly simplified when color is used and the stripes are composed of highly structured patterns. With a strong structuring, our visual system does not have to search long to find a clear assignment of the stripes to each other. You can check this relief of depth perception yourself in Fig. 6.34f.

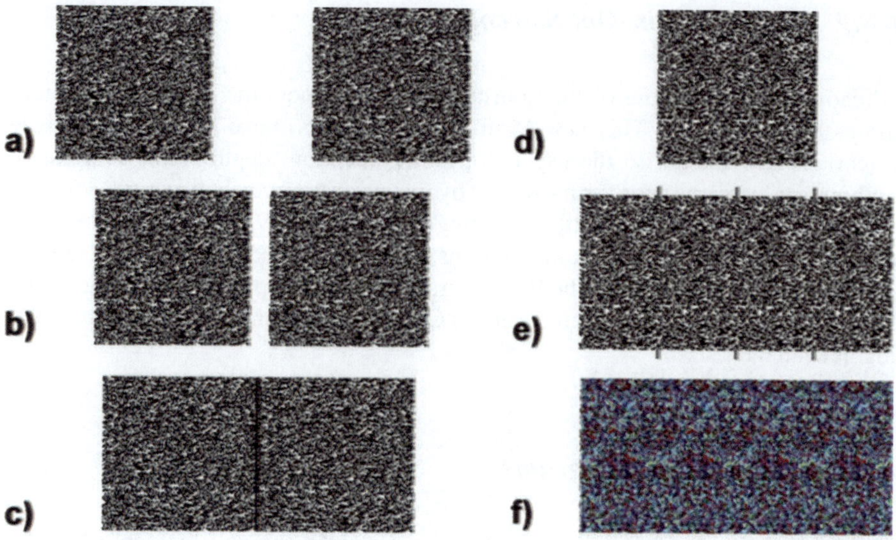

Fig. 6.34 The development stages of autostereograms: (**a**) At the beginning there is a random-dot stereogram pair. (**b**) Depth vision is simplified when the distance of the images become smaller. (**c**) It becomes even better if the image distance disappears completely. (**d**) The image distance can be reduced even further by reducing the image width. (**e**) To restore the image to its original width, we use the wallpaper effect and use a whole series of repeating patterns. (**f**) Stereoscopic viewing is further facilitated if the images are colored and provided with strong structures

The structuring of autostereograms can go as far as the use of completely normal "natural" patterns, photographs, or drawings.

Autostereograms thus ideally combine the advantages of their direct precursors, wallpaper patterns, and random-dot stereograms. The depth can now be perceived with the naked eye just as easily as with the wallpaper effect. In addition, the element of surprise that comes from the random dot patterns is preserved.

It is precisely this mixture of curiosity and fantasy that makes these pictures so appealing. Nowhere else than with autostereograms does the power of vision become so evident. Once the different gaze techniques have been learned, our eyes possess a whole new power.

In the meantime, the current time appears on the display of the time machine. We are back in the present. At the end there are credits running on the display, which briefly summarize the trip through stereo history.

6.8.11 Summary

In retrospect, a constant change in the public interest in stereo vision can be observed. Interestingly, the peak phases of public participation were repeated at an almost constant interval of about 30 years.

The first fashion wave for stereoscopic vision peaked around 1860. There were further crests of waves at the turn of the century as well as in the twenties, fifties, and in a highly commercialized form in the nineties of the last century.

The temporal distance of the waves remained about the same, but not the height of the waves. The wave crests of general interest have risen to ever greater heights due to ever easier dissemination possibilities using new printing techniques and, above all, the further development of stereo imaging techniques!

The final breakthrough of 3D technology now seems to have been the avalanche of good and successful 3D movies triggered by the big Hollywood studios over the last years. 3D is popular and will soon probably be a matter of course in our living rooms in 3D televisions and 3D Internet.

"Hello time machine: What does the future of seeing look like?" asks a fellow traveler.

The answer: "Of course I won't tell! Then everyone would find books like your travel guide boring here. Nobody would publish 3D books and movies anymore, and the whole development would have to start all over again!"

To be on the safe side, we'd rather settle for this oracular explanation before the time machine starts one of the longer explanations about causality, the speed of light, cause and effect, a story about a father who became his own son through time travel, and so on, which are usual at such moments. And all this probably only because it does not want to admit that it cannot go into the future at all!

6.9 New Wonder Worlds of Perception

6.9.1 Multiple Worlds and Ghost Images

Look at the autostereograms with the stereo view in Fig. 6.35. Probably you will discover two normal rabbits in the upper half. Now let your gaze wander to the lower image. It's exactly the same picture, but scaled down. Nevertheless, you probably recognize something completely different, probably a collection of four strangely overlapping rabbits. So there is much more to autostereograms than we have seen so far. With an appropriate look, bizarre ghost images like the doubled rabbits can be seen. The only thing that matters is the depth of view.

In the previous "simple" stereo view, the adjacent structures served as anchor points for the visual rays. See Fig. 6.36 on the left. The remarkable characteristic of the ambiguity of the depth impression has its cause in the frequent repetition of the same picture strip. Now it is also possible that instead of the next but one neighboring structures serve as anchor points for our view. This impression of "double ghost images" is created by "double stereo staring" (Fig. 6.36 center).

Of course, triple (see Fig. 6.36 right side) and even more ghost pictures can also be seen. However, their recognition is becoming increasingly difficult.

Fig. 6.35 Rabbits multiply through different perspectives

The double- and multiple-stereo view, and especially the jumpy transition between these different worlds, is something for advanced users, but in principle it is just as easy to learn as the simple stereo view.

Just look deeper into the picture than in the "simple" world of depth!

Experience shows that at the beginning it is quite difficult to achieve the transition between the worlds if desired. The reason for this lies once again in the good shape of the depth impression already created. Our brain does not see any reason to abandon the once discovered, simple depth perception, even by cautious eye changes. This is why you should make a very extreme change of perspective—at

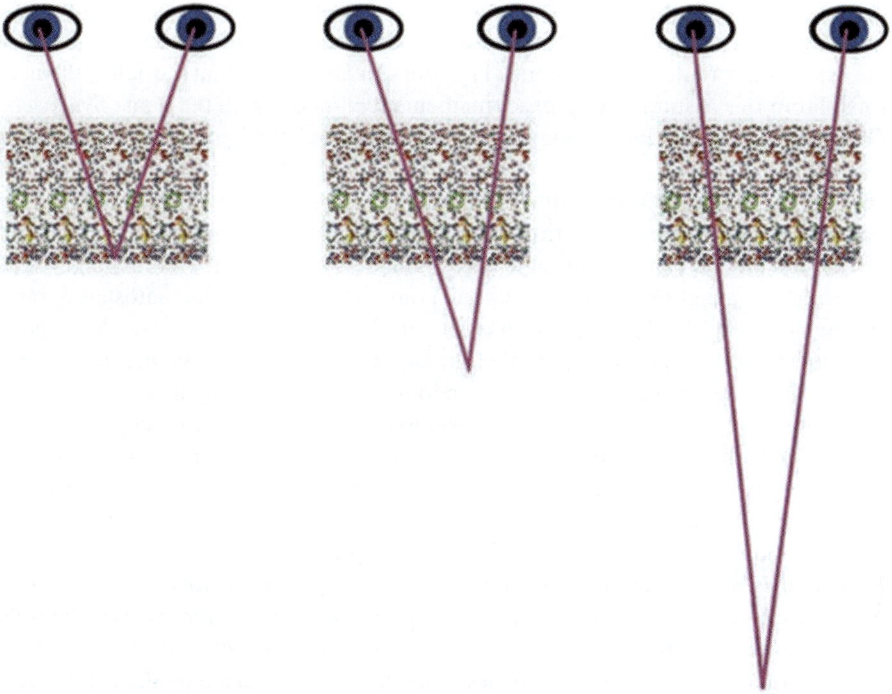

Fig. 6.36 Double- and multiple-stereo view

some point your perception system will no longer be able to withstand this pressure, and at a stroke a new multiple world will appear in front of you! The deliberate transition between the different worlds places a very high demand on our eyes and is certainly only possible in a very advanced stage.

But beginners can do a trick again: As soon as you see the depth image, close your eyes. Now look at a greater depth in your mind only. This significantly weakens the remarkable stability of the previously achieved perception. Now there is a far better chance to reach the transition to other depths as soon as you open your eyes again.

The deeper the multiple world is located, the more the line of sight of our eyes must widen. As soon as eye beams are parallel, the natural limit is reached. This physical limit can be easily exceeded for viewing autostereograms in reading distance. Normally at most double images can just be seen. This situation changes quickly when either the strip width or the whole image size is reduced.

We can summarize: This ambiguity of depth perception is another very special feature of autostereograms, in which many surprises certainly still lie dormant.

In contrast to the pairs of random dot images, many different depths can be perceived here. Once again the autostereograms show themselves as fascinating "quick-change artists" that depend solely on the moods of our senses. Nevertheless, these exotic birds of paradise of perception can be described by science in the meantime.

Synergetics, the theory of the interaction of animate and inanimate individual particles (cf. Journey 4), is again helpful for this purpose. Synergetics assumes that the brain is a complex, self-organized system of many individual particles, the neurons. From this assumption an exact mathematical model with the name *Synergetic Computer* was created in analogy to the human brain. This Synergetic Computer is capable of explaining and predicting a variety of phenomena in psychology of perception, including the recognition of ambiguous patterns, the recognition of human facial expressions, the recognition of handwriting, and depth perception.

The Synergetic Computer is able to recognize all conceivable stereo images such as red-green anaglyph images, random point image pairs, and autostereograms (Reimann et al. 1995). The mechanism of depth-perception of the Synergetic Computer is directly modeled on the human being. Therefore, we can use it very well to illustrate our visual process when looking at autostereograms.

Figure 6.37 shows the temporal development of the depth perception of the Synergetic Computer when looking at the rabbits autostereogram from left to right. The Synergetic Computer is given Fig. 6.35 above with the rabbits, and its reaction is observed with the help of a screen. Different eye positions are specified: single staring, double staring, and squinting. The depth perception of the Synergetic Computer is represented by color tones: Blue means farthest forward, orange-red means backward. The intermediate color tones shown in the color scale on the right represent the corresponding depths between these two extremes. If the Synergetic Computer has not yet found a depth assignment, this is indicated by the color black.

With all three viewing methods, fascinating different landscapes of depths are now gradually growing out of the diffuse starting images. Finally a very exact depth map results. Here the fantastic feature of autostereograms to slip into different roles becomes particularly clear. Depending on the viewing method, the same autostereogram can be used to create various depth maps. In the upper row of Fig. 6.37, the Synergetic Computer has set up the simple stereo view. After some time the two rabbits become visible. The middle row of Fig. 6.37 shows the processes at double stereo view.

The corresponding ghost images of the rabbits become clearly visible on the depth map in the course of time. The lower row shows the squint view. This is exactly the opposite depth effect to the simple stereo view in the upper line!

The result produced by the Synergetic Computer corresponds perfectly with our own depth perception for all three types of vision.

Experiment a little further with the small rabbit stereogram. For example, multiple worlds can also be discovered by squinting. However, this requires an extreme strabismus. The stronger the strabismus, the wider the multiple world appears to "jump out" of the leaf.

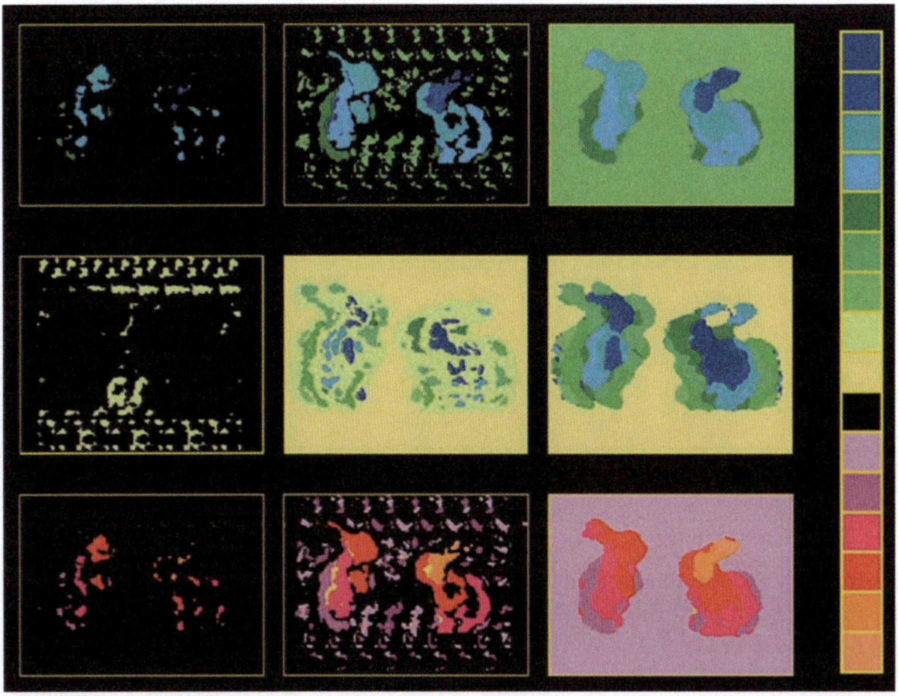

Fig. 6.37 The Synergetic Computer as a model for human perception looks at the rabbits autostereogram of Fig. 6.35: From left to right, the recognition of the depth impression is shown at different points in time. Black means that the computer has not yet recognized anything, blue symbolizes the foreground, and orange-red the background. The intermediate color tones shown in the color scale represent depths in between

6.9.2 An Eye Test to Determine the Depth of Convergence

In the eye test in Fig. 6.38, you can test your dexterity in dealing with the various vision techniques. At the same time, you can test your depth of convergence.

First look at the image with your normal stereo view at a reading distance. Watch the numbers. Which number do you clearly see in a uniform depth behind the sheet?

Evaluation at normal reading distance:

1. Extreme beginner, with this line of vision there is only a minimal decoupling of accommodation and convergence. This layer is even usually a bit difficult for the advanced user, because he is used to a deeper look

2–3. Beginners

4–7. Normal stereo view

8–14. Advanced to extremely good stereo viewer

16. Hardly perceptible at reading distance, as the distance between the numbers is already close to our pupillary distance

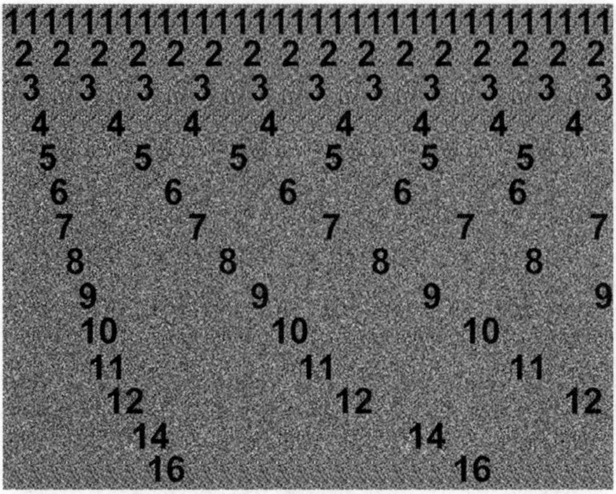

Fig. 6.38 Test image to determine your maximum depth of convergence

If you view the image from a greater distance, you can significantly increase your maximum performance. Conversely, it becomes increasingly difficult to see high numbers in depth as you move the image closer to your eyes.

6.9.3 The Brain Forms its Own Three-Dimensional World

The fantastic ability of our perceptual system to recognize good figures even in depth can be seen in the series of autostereograms with less and less dark dots (Fig. 6.39): While the upper left autostereogram still has a normal dot density with 50% black, the upper right one has only 20% dark dots on the surface. Nevertheless, the depth structure can still be seen very well—even the discontinuous depth jumps, which are feared by all 3D image producers. They disappear when viewing the actually continuous spatial bodies—the 3D objects appear smoother with less surface information. This remarkable achievement of our brain can again be explained by our striving for good form: The brain complements the missing indications of depth due to the lower texture information independently and continuously. In contrast to the autostereogram with a lot of information, where layers are inevitably present due to the print resolution, our perception creates perfect, smooth surfaces in the depth!

Even if the dot density on the texture surface of the autostereogram is further reduced, the depth object is still visible—you can see this in the lower image with a 10% proportion of dark points in Fig. 6.39: The clues on the surface are now quite sparse. Again our brain completes this information to a depth perception. Above a certain minimum point density it is finally impossible for our eyes to recognize any depth. The given information is then simply no longer sufficient to recognize anything.

Fig. 6.39 Even with a very low dot density of the image surface, our perceptual system is still able to detect depth

6.9.4 The Pulling Effect: Our Brain Is Lazy, But Not Too Lazy!

A fantastic perceptual effect already known to us is the *pulling effect*. This effect, originally found on the random dot image pairs (cf. Fig. 6.11), can be wonderfully reproduced when viewing autostereograms. For this purpose, our senses are put to the tęst in a sophisticated way. In the left autostereogram in Fig. 6.40 you can see a square floating above the background in the center of the image. The square was shifted by four pixels, resulting in a clear height above the background. The trick with this autostereogram is that the square can appear to float at different heights, depending on how you look at it. This ambiguity was created by the extreme reduction of the width of the vertical image strip, which now has a width of only eight pixels. Therefore, it is possible to see the square at pixel height 4, 12, 20, 28, and so on.

What depth do we see now? As expected, our brain chooses the simplest option again!

As if the square were a heavy piece of furniture and the brain a furniture mover, it decides for the least effort: The square is lifted from the floor as little as possible.

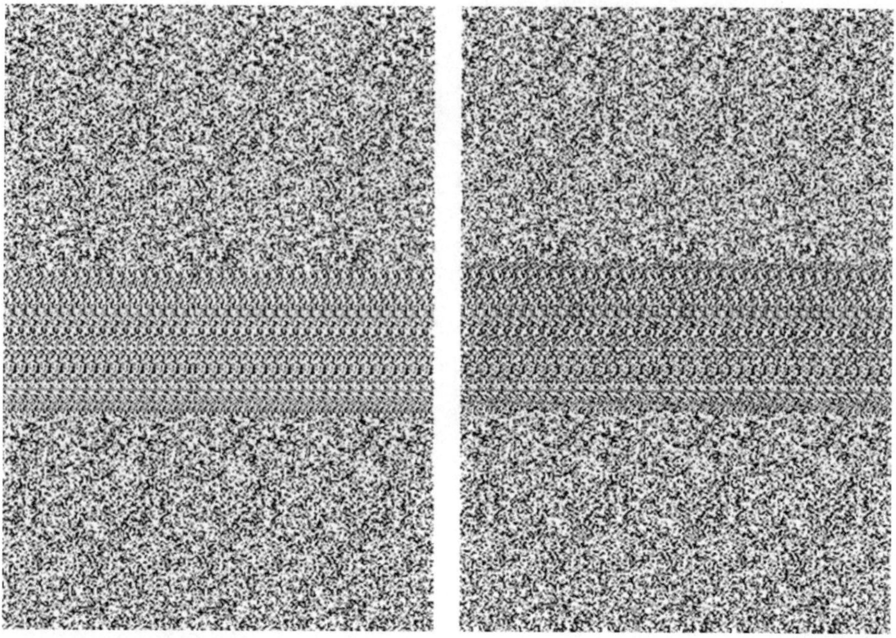

Fig. 6.40 The pulling effect: With these two images you can convince yourself of the laziness of your eyes—but also of the fact that this laziness is not boundless

For comparison, in the left partial picture in Fig. 6.40, a square with the clear pixel height of four pixels can be seen at the top and a square with a height of twelve pixels at the bottom. The square in the middle clearly appears at the same height as the upper square.

What happens now if something is manipulated in the ambiguity of the position of the middle square? This is achieved by the fact that 10% of the pixels in the center of the image have clear depth information. You can carry out this eye experiment in the right-hand partial image in Fig. 6.40—the image point height of twelve pixels was specified, but only 10% of it. This puts our perceptual system in a conflict: either further "lazy," but 10% wrong, or more work and correct. The decision is very clear, as you can see from the depth comparison with the other two floating comparison squares. As before, the upper comparison square is four pixels high, the lower square twelve pixels. The square is now raised by eight pixels in the depth perception and thus drawn into the higher floating state. Therefore the square now appears at the same height as the twelve-pixel-high square at the bottom—this magical effect is again the pulling effect. Fortunately, our perception cannot and will not afford to suppress clear information from "laziness," however small it may be.

6.9.5 *An Eye Test to Determine the Capacity of Depth Perception*

Surely the distinction of the heights of the upper and lower square in Fig. 6.40 was not too difficult for you. But there are also depth layers that are not so easy to distinguish. Our spatial depth resolution capacity depends on the distance between the eyes and the visual acuity, which is determined, among other things, by the density of the photoreceptors on the retina.With the help of an autostereogram an eye test is possible which examines your capacity of spatial perception. Figure 6.41 allows you to check your depth vision yourself. Which of the geometric shapes is at the front? Hexagon or circle? Pentagon or ring? Square or triangle? These questions increase in difficulty. If you can answer the last question, your depth resolution capability can confidently be described as very good. Even if you can only answer the first and second questions correctly, you already have good depth vision. If you have always been wrong, you can still tell yourself that your eyes were just too tired, the light too bright or too dark, the noise around you too loud or too quiet, or the geometric figures in the depths were too scientific or too unscientific.

But you can also simply be satisfied with your normal depth vision. Because you already have a fine depth vision, if you have recognized the geometric figures at all.

Resolution of the visual test: The geometric figures in increasing height above the background are circle, ring, triangle, square, pentagon, hexagon.

Fig. 6.41 An eye test of spatial resolution: Which of the geometric shapes appears farther ahead when staring is used: hexagon or circle? Pentagon or circle with hole? Square or triangle? These questions are becoming increasingly difficult

6.9.6 3D for the Advanced

Additional experiments can be carried out with Fig. 6.41. Tilt the image sideways once by starting to slowly close the book. The 3D worlds immediately appear in a new depth! This is because the lateral tilting makes the repetitive structures seem to be closer to each other than they really are. This means that the eyes don't have to look as deep as they do when the image is not tilted.

The apparent change in the spacing of the periodic structures is also the explanation for the next simple experiment: Bend or wave Fig. 6.41 slightly in front of your eyes. As soon as the picture bends slightly forward or backward at one point, a new depth effect is immediately created. You can even design your own depth landscapes with it.

What happens if you rotate Fig. 6.41 around your viewing direction? As the experiment shows, you will soon no longer recognize anything at all. The reason for this is that the normal autostereograms are designed for horizontal viewing. From an angle of rotation of about 5°, the perception of depth finally becomes impossible.

But there are also pictures that can be rotated without hesitation: These pictures have a special symmetry—best is the point symmetry around the center.

An example of this is the autostereogram in Fig. 6.42. This is the *interference microscope image of* a good old CD-ROM at 500× magnification. If you look at this image very closely with the stereo view, you will clearly see several parallel

Fig. 6.42 Interference microscope image of a CD-ROM at 500× magnification: The stereo view shows defects in the form of trenches that were created during production. Now rotate the CD around its axis and you will see that the trenches remain in their parallelism and depth—albeit in a different orientation and width

trenches. These different depths in the stereo view result from minimal production deviations in the CD. Now slowly rotate the image of the CD around its axis. In fact, the trenches remain even for rotation angles greater than 5°. Although the trenches change their orientation and width at a certain angle of rotation, they still have the same parallel orientation and depth.

Such special autostereograms can also be produced artificially: It is best to use radial symmetry instead of horizontal symmetry. This symmetry requires a completely different depth coding procedure and creates significant difficulties in accurately implementing the depth patterns. What is correctly coded in one direction is unfortunately nullified by the additional coding required in the other directions. Therefore, in the final picture of this journey (Fig. 6.43), we will limit ourselves to an artificially generated autostereogram in which depth can be perceived in two directions:

Take heed of the message that you can see in the depth when you look across, and relax a little more on the last few kilometers of your journey through the world of autostereograms with a look at Fig. 6.43. After all, the longer viewing of an autostereogram is an ideal way to relax.

Autostereograms are therefore more and more used as meditation aids. The component that makes a decisive contribution to relaxation when viewing autostereograms is the new visual technique of stereo vision, i.e., the decoupling of the line of sight of the eyes from the focusing! This technique is completely unusual for our senses and allows our "normal" look a refreshing break, which we have now honestly earned.

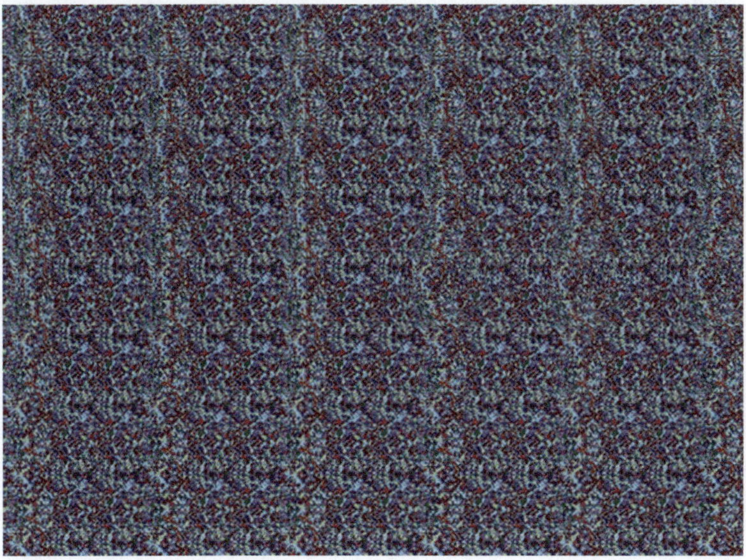

Fig. 6.43 An autostereogram that can be viewed both from below and across

Chapter 7 Seventh Journey: Movements are Life

Ant roads, Tetris stones, clouds, marathon runners, and well-slept toddlers: life is in constant motion!
This journey through the perception of movement leads us into a fast-paced adventure to the limits of research. We encounter—with squeaking wheels—wobbling stars, flickering heat, helicopters, trains, waterfalls, a moving pencil, and rotating snakes. In addition, the simple rotation of a platter straightens the inclined tower of Pisa.

Let the fascination of your vision of motion move you!

7.1 Motion Detection

Our bus driver, elephants, mosquitoes, plants, and bacteria—all life is in motion. Therefore, it is no wonder that the perception of movements is an indispensable part of our success in evolution. The mechanisms of motion perception were established very early in evolution in a rudimentary form. Moving animals were mostly either important prey or dangerous hunters, before whom one had to tear out. Therefore, the rapid recognition of movements was an important key mechanism for survival and a basic prerequisite for visual perception in humans and animals. It even seems that only the vision of the most highly developed animals is capable of transmitting signals to the brain in the absence of movement. For example, a hungry frog can only recognize a fly when it is moving. This is a good indication that motion perception in evolution originated before the ability to recognize patterns in static images.

"If everyone is there now, we can finally get going," the bus driver says after these introductory remarks as all the passengers hurry into the bus, which today is on its way through the world of motion perception. Our driver always likes to take this fast and highly contemporary topic as an opportunity to finally be able to really "put the pedal to the metal."

T. Ditzinger, *Illusions of Seeing*, https://doi.org/10.1007/978-3-030-63635-7_7

He is looking forward to a lively journey with many new and fascinating deceptions. As soon as everyone is seated, he drives up with squeaky wheels and begins to talk about the onboard microphone:

> The rudimentary archetype of motion vision can still be experienced in our highly developed human retina, right at the outermost edges of the retina. These edges are only sensitive to movement. You can convince yourself of this by letting someone wave a certain object at the outer edge of your field of vision. The result: You will recognize the movement as such, but not the object! As soon as the movement stops in this area, the object becomes completely invisible. So this attempt brings us very close to our hidden primitive vision from earlier times. If movements take place farther outside of our field of vision, they are no longer consciously perceived as movement. Instead, there is a kind of unconscious reflex that turns our eyes to the cause of movement and brings it into the central focus. The edge of our retina is therefore—despite the simplest function from the early days of evolution—a kind of early warning system. These rapid warning mechanisms are becoming increasingly important these days, as movements in our high-tech times continue to accelerate in the direction of "faster, higher, farther." We have to find our way in the ever faster road traffic, airplanes and spaceships fly by the air, and Tetris stones, for example, and all other computer games and computing speeds are getting faster and faster.

During this long speech, the bus driver has repeatedly proven, through his driving style, how our road traffic is getting faster and faster and how strongly the early warning system of his motion perception is required by it. But you get used to everything over time.

Our and the bus driver's eyes can pick up the incoming motion information in two different ways.

- Either the direction of the eye remains stationary: the image of the moving object then moves over the retina and generates speed signals there.
- Or the eyes follow the moving object directly: Then the images of the object remain relatively stationary on the retina. Thus, the receptors cannot directly signal the movement, although we can see the movement of the object. When the object moves against a fixed background, we get speed signals from the image of the background now moving across the retina.

It is remarkable that we also perceive movement when the background itself is not visible. This can be illustrated by the following simple experiment: Ask someone to slowly move a burning cigarette in the dark. Follow this with your eyes! The movement of the cigarette is clearly perceived, although no visible background moves over the retina. This experiment shows that the movement of the eyes alone is sufficient to produce a perception of movement.

In the meantime, the bus driver has driven us to a train station to observe relative movements.

7.2 Relative Movements at the Station

The group gets on a commuter train and the bus driver says:

"The perception of movement allows for ambiguous views. You can experience that very well in a train like this."

"How is that supposed to work? In a train you usually only experience delays," says our neighbor.

But the train is already departing.

"I take it all back," says the neighbor, looking out the window at an opposite train.

But after some time, the train opposite is gone—and we are still standing at the station!

Surely everyone has experienced a comparable scene, especially if you have been waiting a long time for the train to depart. This gives the impression that our train is moving in the opposite direction to the train that is slowly starting at the other platform.

This ambiguity is only resolved by motor and haptic perceptions—or even when the other train is no longer standing on the platform.

In trains that are only slowly accelerating at the beginning, the motor information of the movement (through the sense of balance, the skin, or the perception of the inertia of our body) is reduced, and the illusion is thus made possible.

The existence of these two equal perceptual alternatives is conditioned by the relativity of the different reference systems. The optical motion signal that falls on our retina (the *optical flow*) has an ambiguous information content, since only the relative movements between the reference systems (e.g., relative to the other train) are mapped.

In the meantime, our train has already been delayed for a long time and the impression of starting is almost there with the eyes closed.

There is a jolt and our train finally, and conspicuously quickly, starts to move opposite the train standing on the other side of the platform. Again, it takes quite a while before the participants realize that this is also a special case of relative movements: Both trains were travelling—in opposite directions—at the same time. This results in the perception of extremely fast initial acceleration of the departing train!

To perceive movement as realistically as possible, our brain must first decide which reference system is moving and which is stationary. In a self-produced movement such as running or cycling, it is obvious that the optical flow of our environment is clearly caused by our own movement.

If there is no relevant motor information about a relative movement, as in the case of our slowly approaching train, our brain proceeds according to its already well-known pattern of success: It takes a spontaneous decision according to pragmatic aspects.

The Gestalt psychologist Karl Duncker found a simple Gestalt law for relative movements in 1929: If movement is perceived only by seeing, the largest objects are perceived as stationary, whereas the smallest appear to be moving. This is because smaller objects are normally more mobile (and easier to accelerate) than large ones. Figure 7.1 shows a simple experimental setup for such an induced movement. A stationary point of light is projected onto a screen in a slightly darkened room. The special feature is that the whole screen is now set in horizontal oscillations, while the point of light remains fixed.

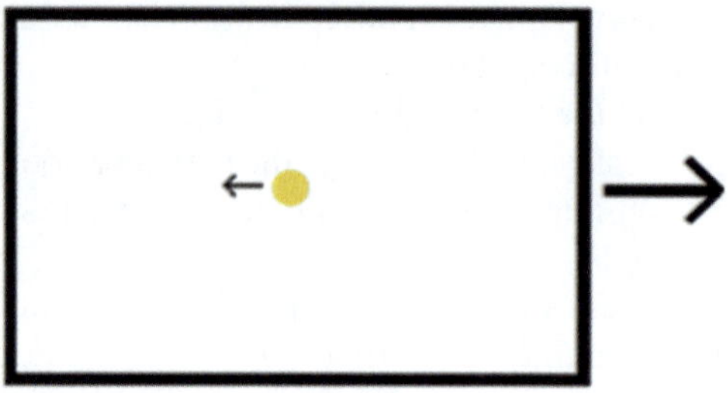

Fig. 7.1 Induced movement after Duncker

Duncker has shown, however, that something completely different is perceived: It looks as if the light spot is moving and the screen is fixed—due to the difference in size between the screen and the light spot! Only by increasing the size of the light spot does the physical information become so compelling that this law of conciseness is overcome, and a moving canvas is visible.

Another example of induced motion is the known apparent motion of the moon through the clouds—although in reality the clouds are moving, and the moon is dormant at this timescale. The reason for this everyday deception is again the law of form of the induced movement, that small objects (the moon) are perceived as movable opposite larger objects (clouds).

After a nice train ride, our travel group has arrived at their destination—an adventure park.

The bus driver explains: "The ambiguity of relative movements is cleverly exploited in theme parks. A strong effect is often already possible with very simple means."

For example, around the castle of Ludwigsburg in Germany there is a remarkable swing, the Herzogschaukel—a kind of ship swing for about 20 people inside an old half-timbered house. After the harmless-looking swing starts to move, comfortably at first, it quickly becomes clear that something is not right, the walls start to tip over, you are accelerated by leaps and bounds and finally overturn several times. Even passengers experienced in amusement parks and looping rides will experience a completely new kind of nausea. Only when you simply close your eyes does the world come back to normal and the swing swings comfortably again.

Solution to the riddle: The entire interior of the half-timbered house begins to rotate relative to the swinging movement after some time! This induces the visual experience of a rollover and this experience clearly overrides the motion perception of the swinging movement, which, in reality, remains comfortable.

Panorama projections (360° cinemas) of films from roller coasters, airplanes, or cars can also considerably disturb the sense of balance.

From such a cinema our group comes out staggering and is happy to listen to the bus driver standing still:

> Another example of ambiguous relative movements is the famous seasickness, which usually occurs inside a swaying cabin. The problem again is that the motor and visual motion detection are not in harmony. While vision gives us the impression of a fixed, stable reference system, motor perception points to strong accelerations.

Only when the seasick person steps on the railing (attention: headwind) and the horizon is visible, the optical and motor perception of movement are in harmony again and the nausea usually subsides.

A fellow traveler says, "There is a similar phenomenon when driving a car. When I read in the car during a winding drive, my stomach soon grumbles. Only when I look out the window for a while does the nausea usually calm down again."

"I've never felt sick while cornering," says the bus driver—and everyone is glad that they're not on the bus right now.

The bus driver would like to lead the group into a mobile high-tech cinema cabin in the shape of a small airplane, which moves jerkily with an action film shown. But everyone refuses with thanks and looks forward to the screening of the illusory movements—because there they themselves are no longer moved.

7.3 Illusory Movements, Movies and the Wobbly Pencils

The first illusory movements were demonstrated by Exner in 1887. He generated two electrical spark discharges in short time intervals from each other at different places. The impression of a movement between the two spark locations was visible! As the Gestalt psychologist Max Wertheimer showed in 1912, the nature of the resulting apparent movements depends on the spatial and temporal distance between the two light pulses.

Thus, an optimal continuous apparent motion (also known as Phi motion or stroboscopic effect) from one light source to the other is observed for time intervals of 60–200 ms at a spatial distance within the normal visual range.

If, for example, the two circles in Fig. 7.2 above are alternately illuminated on a screen at intervals of 100 ms, a circle is perceived that moves continuously back and forth! These illusory movements can again be understood with a well-known recipe for success of our perception: the striving for the simplest and clearest possible solution. And this form of movement is that of a continuous movement between the two light sources.

The desire for a simple form of movement is so strong that other laws of seeing are simply "overruled." For example, during the apparent movement, forms are transformed into one another without hesitation. As shown in Fig. 7.2 below, a circle is transformed into a square during the perception of the apparent movement, and vice versa. The perception of movement thus dominates the perception of form here.

The stroboscopic effect can be used excellently for the investigation of different Gestalt laws. A simple arrangement, as shown in Fig. 7.3, is suitable for this. Four alternately illuminated circles are arranged in a square.

Opposite circles are illuminated at the same time. This creates an ambiguous movement stimulus. Horizontal or vertical oscillations can be detected, as well as ring-shaped clockwise or counterclockwise oscillations. Between these different patterns of movement there are many perceptual transitions in the form of the

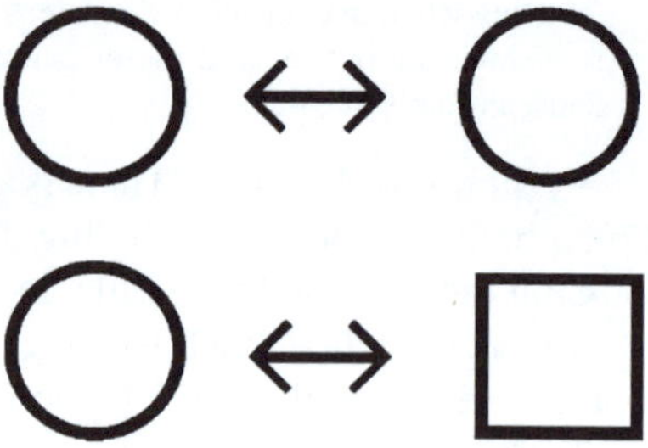

Fig. 7.2 Simple illusory movements

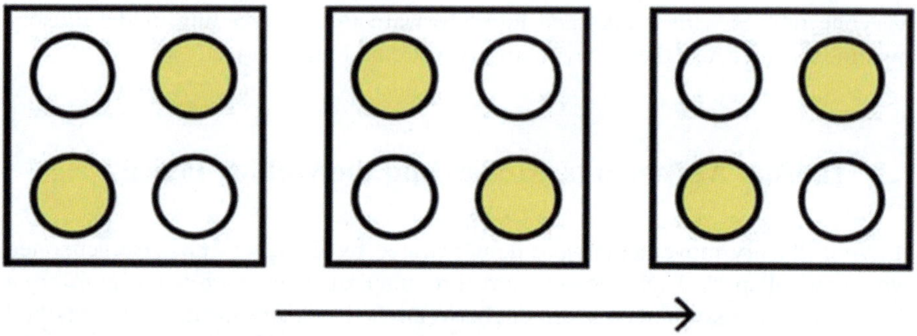

Fig. 7.3 Apparent movements, the time runs from left to right

oscillations of perception, which were described in Journey 4. The weighting of the alternatives can now be changed simply by applying different laws of form. For example, if the vertical distance between the circles is increased, the horizontal oscillation is preferred due to the law of proximity, and so on.

Probably the best-known apparent movement can be admired every day in cinema and television. Although everyone knows that cinema and television consist of a sequence of still images, continuous sequences can be seen. The explanation for this is again the illusory movements and the *persistence of perception*. The persistence (temporal inertia) of perception is the inability of the retina to follow rapidly changing intensities. Flickering light, which illuminates more than approximately 50 times/s, appears to be steady under normal lighting conditions. The frame rate of films for television and cinemas has been 24 frames/s for almost 100 years. Therefore, unfortunately, it does not always induce a completely flicker-free perception, especially with medium-fast object movements. This is also repeatedly criticized, for example by star director David Cameron, who originally wanted to shoot his 3D film *Avatar* in 48 frames/s, but ultimately failed for cost reasons.

It is well known that conventional television pictures are not displayed as a complete overall picture. Rather, the electron beam involved builds up the image line-by-line from top to bottom. This special technique of spatial image representation is, of course, only made possible by the persistence of our perception.

A fascinating example of a comparable but spatially reversed process, also based on persistence, can be found at the Epcot Center in Disneyworld Orlando. As in

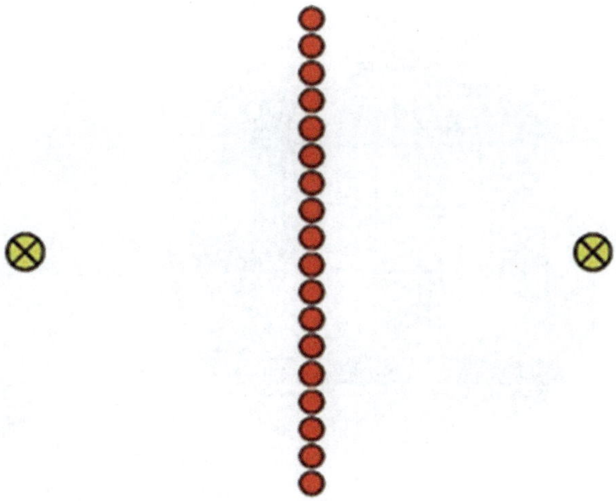

Fig. 7.4 Images from nowhere: deception through guided eye movement

Fig. 7.4, a single vertical column equipped with flickering light-emitting diodes can be seen on a wall in the middle. On the left and right side there is also a single light emitting diode, which light up alternatingly.

This arrangement only makes sense if you follow the instruction to alternately fix the currently lit diode at the edge.

The vertical light-emitting diodes are imaged at different places on our retina by this given change of viewing direction of our eyes. The flickering of this chain of lights is precisely timed to the given horizontal eye movement, so that real images emerge from nowhere that seem to extend across the entire wall surface! This fascinating new, active way of watching television is unfortunately not for the mass market, as it quickly becomes too exhausting for the eyes.

"Yes, when watching television, just opening the chip bag and pressing the remote control is enough for me to make an effort," says one participant.

"And to see how the cyclists toil in the Tour de France is activity enough for me, I really suffer there," says another participant.

"In fact, nowadays you sometimes feel as if you are sitting on your bike in the middle of the action, because watching fictitious movements in the cinema or on television can normally no longer be distinguished from real movements," says the bus driver.

But there are exceptions!

The classic case is the wagon wheels of a carriage, which appear in every good western and seem to turn backward. This wagon-wheel effect is physically justified: The wheel rotates at comparable or higher rotation rates than the film's image sequence. If the rotation of the wheel is approximately slightly slower than the frame frequency, the snapshot of a wheel that is minimally backward is taken in each partial frame—which in the end physically means a wheel turning backward. Since the wheels of carriages have many spokes and thus a particularly fine symmetry of rotation, the wagon-wheel phenomenon already starts at much slower

Fig. 7.5 Persistence of perception: This is how the bird gets back into the cage: Please stick the two pictures together and turn them around quickly!

frequencies. For example, with 20 spokes, the wheel can turn 20 times slower than the sequence of images in the film to show the observed effect. This brings the wheel rotation rate into the range of about two revolutions per second.

Another well-known example is starting a helicopter or propeller plane. The cinematic view of the accelerated rotor or propeller lets us viewers walk through a whole cascade of forward and backward rotating propeller interferences. This phenomenon has become so internalized in us that we no longer wonder about it. Rather, we may be surprised during an airport visit when we see this scene in reality for once and the movement is continuous!

The difference between illusory movements and persistence can be easily understood by looking at the bird that flew away and the cage in Fig. 7.5.

How does the bird get back in the cage?

The solution to this problem goes to an English physician with a French surname, John Ayrton Paris. He developed a toy he called a thaumatrope as early as 1826. Cut out a photocopy of the two pictures, stick the two back to back on a cardboard and bring the whole thing to a fast rotation. This can be done, for example, by using two rubber bands that are attached to the left and right of the picture and unscrewed. The bird sits now clearly again in its cage and that although always only one of the two pictures is seen. The cause of the deception is again the persistence of perception between these two different images which alternate in our field of vision – with a frequency too high for us to physically resolve them in time. An illusory movement, on the other hand, is not involved here. Illusory movements come into play only when it comes to the recognition of similar, successive image sequences, such as in a flip-book cinema.

A wonderful illusion based on apparent movements is the effect of the wobbly pencil, well-explained by e.g. Pomerantz (1983), in which a pencil, as in Fig. 7.6, is set into a loose finger position and vibrates.

The bus driver is demonstrating this deception to the tour group right now.

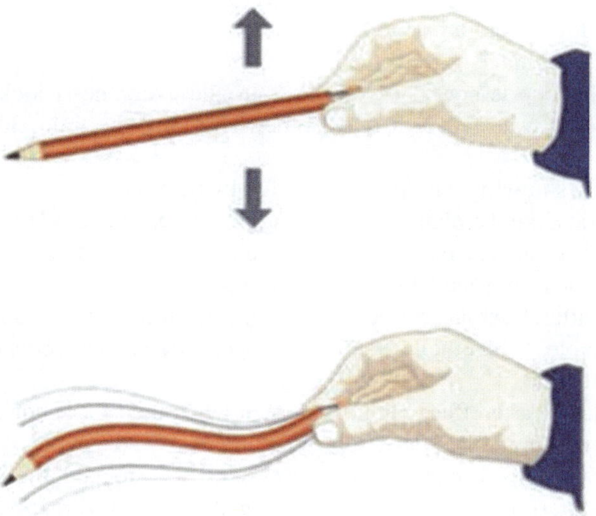

Fig. 7.6 A movable rubber pencil

"Looks like the pencil's made of rubber," says our neighbor.

"Yeah, like when it swings all by itself," says another fellow traveler.

The bus driver is very happy that he has succeeded in this trick—he has exercised it for a long time.

The trick is this: Hold the pencil close to one end and very loosely between thumb and index finger. Move your hand vertically up and down in short, fast movements. The pencil should not be deflected by more than 5 cm. Especially important for the trick is that the whole hand swings in phase with the pencil deflection. This creates a nodal point (standstill) of the pencil movement approximately in the middle of the pencil—clearly away from the grip of the hand on the pencil.

The cause of deception is the experience of elasticity next to the apparent movement.

The fact that a pencil moves strongly at one end, not in the middle, and then in counterphase, can—contrary to everyday experience—be interpreted most simply by "elasticity/pencil is made of rubber."

7.4 Aftereffects, Waterfalls and Trains Again

At the next opportunity, look at a waterfall for some time (30–60 s is enough). If you then look at any other still scene, you will notice an exactly opposite movement as an afterimage. By the way, this famous waterfall illusion was already known to Aristotle.

"Yes and to me too," says a fellow traveler, who tells of a similar observation: He once looked down from a bridge at a fast-flowing river and then had the opposite current as an afterimage of movement.

Our travel group is now back on the train and everyone is looking out of the window to see a waterfall or a river somewhere until the train stops at the next station.

"Something's moving with me," says a fellow traveler.

"Yes, it looks like the platform is moving in the opposite direction."

The bus driver is very pleased about the discovery of this very clear aftereffect movement at the right time—that is modern didactics!

A similar aftereffect can be observed in a car, on a boat, or with the help of a rotating turntable. When the turntable is stopped, the platter seems to rotate backward for some time.

A clear motion aftereffect can also be seen when the spiral shown in Fig. 7.7 is rotated on an old turntable or a wooden spinning top. On the platter (clockwise rotation) the spiral seems to expand. Once the rotation is stopped, the spiral seems to contract. The afterimage is thus visible again in the opposite direction of movement.

These fascinating motion aftereffects are proof that motion vision is an independent sensory perception with its own directional motion detectors. The motion aftereffect is therefore comparable to the after-images of color vision we saw earlier: If one observes a certain color for some time, the color receptors involved become saturated and at some point only the complementary color is perceived. In the same way, directional motion detectors on our retina seem to saturate if the motion only runs long enough in the same direction. If one then looks at a still image, the complementary movement, i.e., the movement in the opposite direction, is perceived.

The existence of directional motion detectors is not as trivial as it sounds at first. It would also be conceivable that the perception of movements and velocities could be derived from the internal conversion of the time and location information of the individual photoreceptors. Due to the direction-dependent motion aftereffect, however, it becomes clear that there is a completely independent motion sensation without a change of location taking place.

Fig. 7.7 With clockwise rotation the spiral seems to expand, after standstill an aftereffect of movement occurs in the opposite direction

The key to the organization of these motion detectors on our retina was found by Reichardt (1957) from experiments with insects—for example, the housefly—in the form of sophisticated coupling and correlation mechanisms of two locally separated retinal sites. This model was later extended by van Santen and Sperling (1985) to a general connection of many different retinal sites to a so-called Reichardt detector. The physiological existence of direction-dependent motion detectors was proven by Hubel and Wiesel (1959).

7.5 The Autokinetic Effect and Star Movements

A single light in the darkness always seems to wander back and forth somehow restlessly. This effect was already known to Alexander von Humboldt. He reported in 1851 that a single star appears to move. Schweizer (1857) observed individual differences in the reports on star movement and concluded that this effect is a subjective impression of movement, which Aubert called *autokinesis* in 1887.

The autokinetic effect is best seen in a clear night sky with a single star. After some time, this star seems to move and at times appears like an airplane. However, this imaginary plane flies like a moth around the light around the physical resting position of the star.

There has been much discussion about the autokinetic effect, and to this day there is no consensus on an explanation for this fascinating effect. Explanatory approaches include the movements of suspended particles in the vitreous body of the eye, saccadic eye movements, natural fluctuations, or signs of fatigue. Perhaps the most convincing explanation is the inclusion of the correction mechanisms of the visual muscles according to Richard Gregory (1997). These dynamic correction processes enable us to fix the desired section in a naturally complex, moving environment. In our sensitive, dark, static, and clueless star environment, however, these dynamic correction mechanisms simply overreact and repeatedly overshoot the mark.

As always, when very sensitive and deliberately minimally controllable causes achieve recognizable effects, fascinating theories about the influence of subconscious inspirations emerge. The autokinetic movement of a light spot in a darkened room has already been used as a personality test, and the influence of group dynamic behavior has been noticed. The psychologists Rechtschaffen and Mednick carried out a wonderful experiment in 1955. The subjects were informed that it was an experiment to test the ability to read words written with the light spot. In reality, the point of light stood completely still. The result was that all test subjects observed words—a nice proof that inspiration actually influences perception. Usually simple words were observed, but some test subjects also recognized really long messages. One subject even deciphered a multitude of very personal sentences until she finally asked the experimenters: "Where did you get all that information about me?"

In the meantime, our group has returned safely in the bus.

"All the movements you have experienced so far were just a foretaste of the fantastic movement illusions to come," says the bus driver, and he presses the accelerator pedal with joyful determination.

We let ourselves sink into our seats with suspicion.

7.6 Motion Illusions with Periodic Patterns

"Images with repetitive patterns often create curious perceptual effects—both in the direct view and in the afterimage," says the bus driver.

See Fig. 7.8. In this lively picture after MacKay (1957), moving shadows forming rosettes can be seen after only a short viewing. If you fix the center of the figure, you will notice a distinct rotation of the shadow. The direction of rotation can be reversed by fixing a point to the right or left of the center.

If you move Fig. 7.8 horizontally back and forth, you will notice blurred eights moving at right angles to the real direction of movement.

This fantastic mobility of repetitive patterns is analogous to the famous Moiré patterns. These patterns can be seen when two identical images of certain repetitive geometries overlap. Moiré patterns are always created when the two copies do not lie exactly on top of each other. For example, if you copy Fig. 7.8 onto a transparent foil and do not place it exactly on the original, the perceived ring-shaped shadows will result.

Fig. 7.8 The MacKay image

But how can these Moiré patterns arise in our perception? One explanatory approach is that the rapid involuntary saccadic eye movements are the cause. In a very short period, these movements cause the image of the pattern on the retina to appear slightly displaced. Due to the persistence of perception, these two images interact for a short time—just like the bird and the cage in Fig. 7.5.

Also, the afterimage effect of periodic images is very remarkable. For a demonstration, look at Fig. 7.8 for approx. 30 s and then look at a static monochrome area. You will notice a moving aftereffect, which MacKay called *complementary afterimage effect*. The movement of the aftereffect is perpendicular to the main lines of the picture, so here it is ring-shaped (see also the spiral in Fig. 7.7).

Another beautiful example of the impressive mobility of repetitive patterns can be seen in Fig. 7.9 "Chrysanthemum." The image virtually dissolves in front of the eyes and begins to shimmer and move radially to a great extent. However, the apparent direction of movement of the image is not uniform but depends on the respective orientation of the patterns. This results in various circular rings that seem to rotate clockwise and counterclockwise—and that even shimmer and move at different speeds.

This wonderful picture was originally created by the English psychologist and artist Nicholas Wade in 1982, and an improved version was made in 1990.

"In the meantime, we have arrived at the extreme limits of our perception of movement and our journey," says the bus driver.

Take a look at the whirling vortex in Fig. 7.10. This intense illusion of movement was discovered by the Japanese visual scientist and illusion artist Akiyoshi Kitaoka

Fig. 7.9 Chrysanthemum. (With the kind permission of Nicholas Wade)

Fig. 7.10 Whirling vortex. (With the kind permission of Akiyoshi Kitaoka)

and consists of a very special combination of deceptive mechanisms of spatial depth and movement.

The image pulsates back and forth—so extreme that it is reminiscent of the effect of alcohol at the end of an equally extreme evening. The cause of this effect is not yet fully understood.

This is also similar to alcohol: "The cause of my drinking is sometimes also not fully understood," says a fellow traveler.

In any case, spatial regularity again plays a role. However, this alone is not enough to produce this irritating pulsation effect. For the deception to take place to this extent, the small colored dots must be sharp—and colored.

7.7 Illusions of Movement with Colors

"Based on the Benham disks, we have seen that movement can become color," says the bus driver.

"Is there also the reverse process, that colors can become movement?" asks a fellow traveler.

"The answer is yes. I would like to demonstrate this with a beautiful magic trick in which the Leaning Tower of Pisa is straightened."

7.7.1 The Leaning Tower of Pisa is Being Straightened

All you need for this magic trick, which can easily be done at home, is an old record player, copier and scissors.

Place a circularly cut color copy of the strangely colored Leaning Tower of Pisa from Fig. 7.11 on a platter. As soon as the turntable turns, the tower rises vertically—very beneficial for the statics, but bad for tourism in Pisa.

This fascinating counterpart to the Fechner-Benham illusion was discovered as the "Leaning Tower of Pisa Effect" (Ditzinger et al. 2000). In addition to the inclined tower, all other shapes can also be rotated against the actual direction of rotation—provided the combination of the colors involved is correct.

For example, if you place Fig. 7.12 on the turntable, the green bar is rotated to an orientation that appears parallel to the other blue bars. In contrast, the yellow bar remains in its angle.

Here is green in special correspondence with the red background.

An attempt to explain this new phenomenon—as with the effect of the Fluttering Hearts in Journey 5—is based on the different processing times until the differently colored surfaces are detected. The respective detection times are determined by the combination of color, saturation, and brightness involved. So we already know that blue is perceived the slowest, red and green about equally fast, and yellow the fastest. And we know that light tones are perceived faster and dark ones are registered more slowly. It therefore depends on the whole mix of color, saturation, and brightness.

Fig. 7.11 The Leaning Tower of Pisa is being straightened

Fig. 7.12 The green bar is oriented parallel to the blue bar, while the yellow bar remains constant upon clockwise rotation

Assertion: If this mix is correct and the recognition times for two enclosing surfaces are identical, the mobility of the enclosed surface begins—the two surfaces interact with each other. This condition is fulfilled for the Leaning Tower and background with the colors magenta and cyan in Fig. 7.11. All other color combinations in the figure have different perceptual processing times. As with the Pulfrich effect, this leads to a different apparent position of the various surfaces. The previously statically coherent image becomes inconsistent through movement and falls out into its individual parts. Even more than that: the individual surfaces exactly overlap in the optical flow on the retina.

This creates optical overlays in some places and completely empty areas in others.

Once again, our brain has an urgent need for action! Our perceptual system does not allow a previously consistent image to simply fall apart through movement without necessity. For this reason, the physical visual information is overruled by a Gestalt law of the common good fate of motion. The position of the individual surfaces is therefore simply perceived as fixed to the background!

A special case is now, of course, the case of interacting surfaces, i.e., when their colors are perceived equally quickly. Then there is no sufficient need for action for our perception and the scene is still perceived as spatially consistent. The lack of shape correction allows a deep insight into the temporal visual processing: The enclosed surface is now also recognized as spatially shifted according to the temporal recognition delay! When turning clockwise, the tower is thus simply moved counterclockwise on the blade! The constant stable rotation makes this effect very robust and, in contrast to the similar effect of fluttering hearts (see Journey 5), it also works in daylight and with central vision. Via the perceived angle of rotation, the Leaning Tower of Pisa Illusion even allows a quantitative measurement of recognition processing times.

7.7.2 The Law of Common Destiny of Motion

The law of the common fate of movement can be convincingly demonstrated by extending the effect of the fluttering hearts. Please move the two partial images of Fig. 7.13 slowly back and forth in dim lighting (e.g., candlelight). In the upper line, both fonts appear mobile. In the lower half, nothing moves anymore, although the right half of the picture, including the font, is completely identical to the upper drawing file! The explanation of this enchanting and still undescribed effect is the law of the common fate of movement. As we already know, the mobility of the left text in the upper picture is caused by the effect of the fluttering hearts. Since our perception always strives for a consistent, simple movement pattern, the movement of the right-hand text is simply adapted and a common movement of the two writings is recognized! The left text in the upper picture is the drive for the mobility of the right-hand side text via the law of good movement. This drive is omitted in the picture below, and therefore the right font does not swing anymore. Abracadabra!

"I think everyone has now seen that the perception of colors and movement is directly related," says the bus driver. He continues, "Motion vision is also strongly related to spatial vision—we already know this from the Pulfrich effect, for example." The bus driver switches on the radio and looks forward to his next magic trick to create spatial depth.

Fig. 7.13 Please shake: The Law of Common Destiny

7.8 Illusions of Movement Through Spatial Interaction

If you place Fig. 7.14 on the left with the eccentric circles on the record player, you can immediately experience a spatial perception. The circles protrude far into the depth during rotation in the form of a tube!

This fascinating and unfortunately relatively unknown illusion, which shows the direct connection between space and movement, goes back to the psychologists Benussi and Musatti and was called *stereokinetic effect* by Musatti (1924).

The history of this effect is almost as fascinating as the effect itself and is marked by recurring discovery and forgetting. Thus, around 1935, these pictures regained prominence in an embellished and expanded form as "Rotorelief" by the artist Marcel Duchamp—who was also interested in the fluttering hearts, by the way. He was so enthusiastic about his newly discovered kinetic works of art that he finally presented them in 1935 at a fair for inventors and household goods (the Paris Inventors Fair). He had a small stand of 3 m^2 where he presented his Rotoreliefs squeezed between the presentation of a new type of garbage compactor and a new vegetable shredder on a record player. The performance was a great flop, and his useless invention went practically unnoticed. Instead of payments into his account, his appearance went at least down in art history—as an example of breadless art.

The spatial effect of these rotating images has physical causes. Our perception proceeds analogously to the observation of a linear movement like the flow of a river: river sections that are closer to the observer appear to move faster than flows on the opposite bank. Conversely, conclusions can be drawn about the spatial depth from the perceived movement: The faster the movement of an object, the more it is perceived in the foreground.

Exactly this characteristic in connection with the Gestalt law of conciseness seems to be the reason for the spatial effect in stereokinetic images.

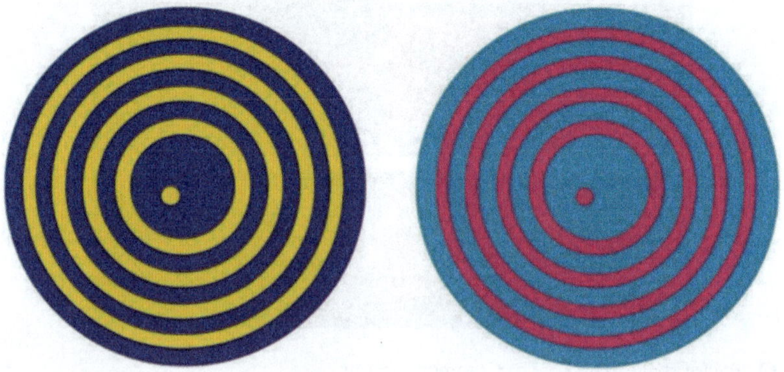

Fig. 7.14 The stereokinetic effect—this image becomes spatial as soon it is rotated, for example, on a record player. (According to Musatti 1924). Picture right: the same picture in interacting colors: The effect fades

The circles as a figure are so concise that our perception tries to maintain them even during movement. The circles are not concentric, so the individual sections show different physical velocities during rotation. Since it is essential to preserve the perception of the good circular shape, our perception here simply invents a new ingenious compromise: the impression of spatial depth! Our perception system again proves to be extremely inventive and makes pragmatic use of all available degrees of freedom. The farther the lines are from the center, the faster they move and (as with flow) the closer they appear to the viewer. This creates the wonderful depth impression of a tube.

What happens when less concise shapes are given as circles? In 1924, Musatti and Benussi experienced (Musatti 1924), instead of spatial depth, a deformation of the figures in the form of a drift around the figure axes for ellipses arranged eccentrically around the center. Our perception again follows its very own recipes for success and goes back to the easiest possible way. There is no longer any need to be considerate of good form. Therefore, the shape is simply adjusted according to the speeds in a temporal wobble. An additional spatial depth does not have to be felt as a "way out" anymore.

"Now we have seen that color is related to movement and spatial depth is related to movement. Shouldn't a combination of all of them create a lot more confusion?" asks one traveler.

"Yes, that's true, but I'd like to save this mess for the end of our trip. Fortunately, sometimes there are also interactions between movement, colors, and depth vision that reduce confusion," answers the bus driver.

Since you already have the turntable in use, we want to check this with the help of Fig. 7.14 on the right. This is the same disc as on the left, but in interactive colors. Let a color copy of it rotate on the turntable. The result is that the two effects cancel each other out and there is no, or only very weak, depth perception. This (previously unpublished) effect can now be easily explained with our findings. Due to the inclined tower of Pisa effect, the eccentric circles are simply shifted by a certain angle against the direction of rotation. The good shape of the circles simply remains and the way out over the third dimension is not necessary—and is therefore not realized!

7.9 A New Fascination: The Modern Illusions of Movement Under the Influence of Color, Depth, Form, and Brightness

A whole new class of fantastic movement illusions has recently been created. Due to the intensive effect on the observer, the playful is often in the foreground—the explanatory approaches, on the other hand, are still in their infancy.

"I had promised that things would get more and more confused in the end," says the bus driver to his group. "So please hold on tight, attention—danger of vertigo!"

7.9.1 The Ouchi Illusion

Look at Fig. 7.15, which goes back to the Japanese artist Hajime Ouchi from 1977 and consists of a simple arrangement of vertical and horizontal stripes. If you change the viewing angle or move the image slowly horizontally or vertically, the middle part of the image gains a fantastic agility. It emerges spatially and seems to move in the opposite direction to the background movement.

The cause of this intense deception of movement is the vertical orientation of the stripe patterns and the perception of spatial depth.

Random eye movements play a central role in this. The special geometry of the two stripe patterns eliminates the perception effect of eye movements parallel to the respective patterns. Thus, only the movement information of the eye movement oriented perpendicular to the stripes arrives in our visual center—which leads to the completely independent movement impression of the two surfaces.

This effect is further enhanced in the Ouchi figure by the circular spatial limitation of the horizontal stripes. Like looking through a keyhole, we perceive the line patterns within the circles in a different spatial depth and movement. This process is a substitute mechanism of our perception for the so-called opening problem. It is well known that the movement information that physically arrives at our retina when viewing scenes through openings like a keyhole is very small. This is why our perception invented replacement mechanisms that make the opening contents appear to have a different depth and mobility. These mechanisms significantly enhance the effect of the Ouchi illusion.

An almost unbelievable new illusion goes back to the Japanese Akiyoshi Kitaoka. He has superimposed a new special blurred texture on the basic circular pattern of the Ouchi illusion, creating a strong additional spatial perspective. As we have already seen, blurred patterns are perceived as lying deeper in the background. This further enhances the keyhole effect and makes it possible to perceive the extreme

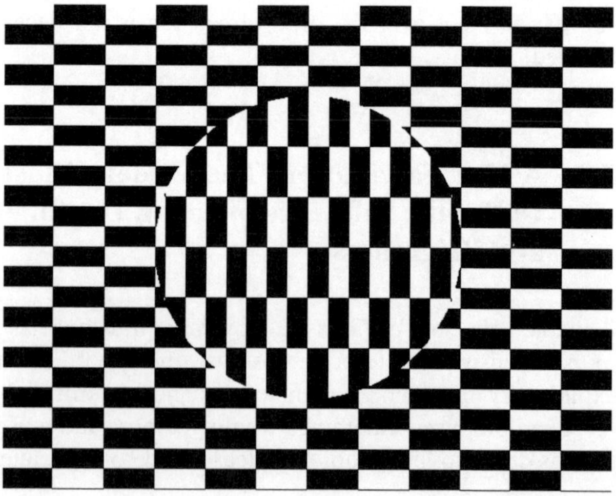

Fig. 7.15 The Ouchi illusion

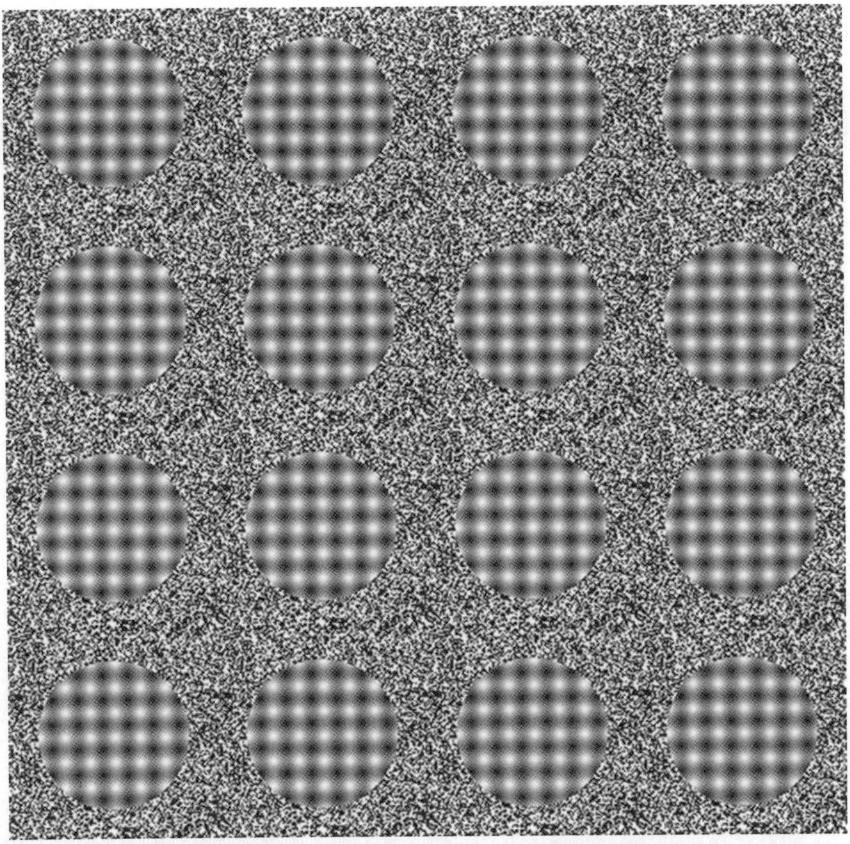

Fig. 7.16 Intensive illusion of movement. (According to Akiyoshi Kitaoka)

mobility of the entire image. The movement effect of Kitaoka's figure can be additionally enhanced by making the openings smaller and appear several times. You can observe the resulting extremely strong motion impression in Fig. 7.16.

7.9.2 Pinna-Brelstaff Illusion

Another movement illusion, which has become known in a short period of time mainly via the Internet, can be seen in Fig. 7.17. It shows the Pinna-Brelstaff illusion of the rotating circles (Pinna and Brelstaff 2000). This figure belongs to the most impressive and strongest movement deceptions—convince yourself!

Fix on the point in the center of the image. Slowly move your head toward the image and then move it back again.

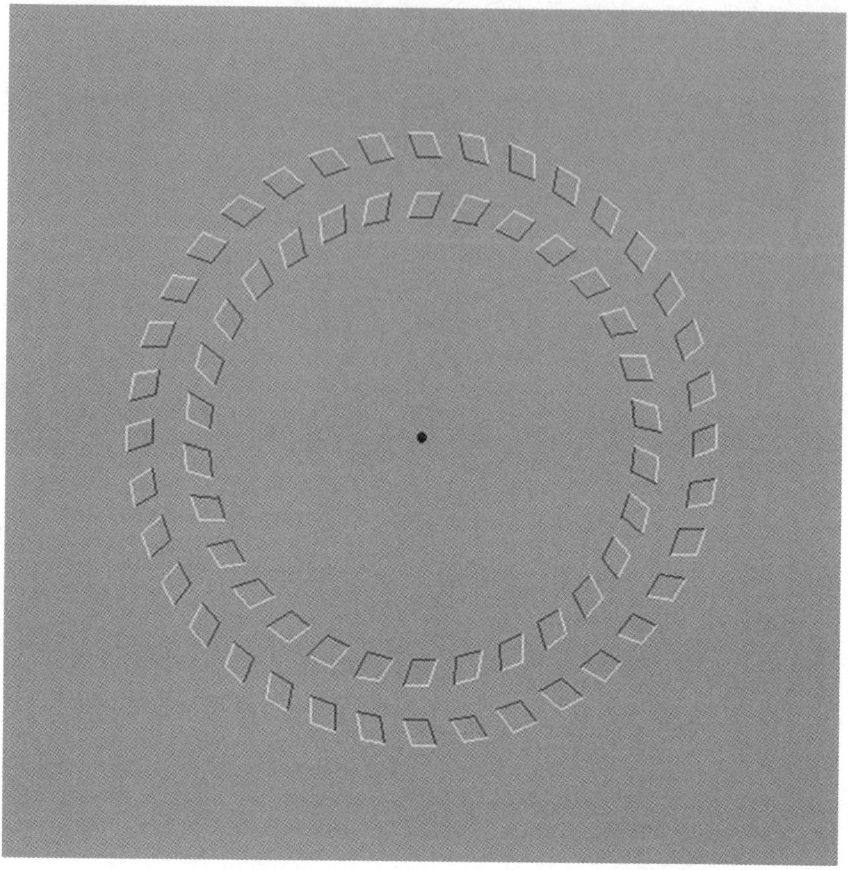

Fig. 7.17 Pinna-Brelstaff illusion of rotating circles. (Courtesy of Baingio Pinna)

Abracadabra—the circles begin to rotate in opposite directions as if by magic! The directions of movement are reversed each time you reverse the direction of your head's movement.

The explanation of why these flat static patterns suddenly appear to be mobile has not been finally clarified. When you move your head toward the sheet, the circles in the sheet plane rotate depending on the orientation of the patterns. Besides the different shape of the patterns within the circles, the different shading at the edges of the patterns plays an important role. Thus, the circles seem to lie at different spatial depths, which are reinterpreted into different movements.

A similar deception can be seen in Fig. 7.18. This is a rectangular variation, and in some way a generalization of the Pinna-Brelstaff illusion.

Even in the static state, you can immediately see that there is something wrong with the image, that somehow the columns seem to want to twist against each other.

Now slowly move Fig. 7.18 vertically up and down. Surely you will see an apparent movement of the vertical columns in a horizontal direction: The columns move

Fig. 7.18 Moving lines. (With the kind permission of Baingio Pinna)

toward each other in pairs! If the direction of movement is reversed, the column movements are reversed.

The trick also works in the other direction: Move the image horizontally to the left and right. Now the lines move in pairs perpendicular to each other.

What is behind this deception? The shapes of the individual static patterns this time are uniformly oriented, completely identical squares.

The only difference between the patterns of the columns is the different brightness of the edges. In columns 1, 3, and 5, the edges on the right and top are light, while in columns 2, 4, and 6, the edges on the left and top are light.

The squares all appear to be spatially elevated in relation to the background, but always with the impression that something is wrong with the picture. We can explain this from our knowledge of the influence of the sunlight coming from above.

In the odd columns the impression is created that the sun comes from the top right, while in the even columns it shines from the top left.

Since our perception has grown up with only one sun during evolution, this image contradicts our everyday experience and seems to be "under strong tension." In this fascinating picture our perception does not find a conclusive solution to make the impression of a double sun disappear without a doubt. That's why our brain just starts cheating: Our perception seems to try to illusorily twist the orientation of the squares as much as possible in the direction of a single virtual sun shining vertically from above. This creates the impression that the columns are trying to tilt against each other.

Fig. 7.19 Movable rectangles. (Courtesy of Akiyoshi Kitaoka)

This asymmetry of the squares, given by the different shading of the edges and their perceived different orientation, also seems to be sufficient as a cause for the strong illusion of movement.

The different shading of the image becomes apparent when you rotate the image 90° clockwise. Surely you will immediately see a different spatial depth of the rows. The first, third, and fifth rows seem to be clearly behind the leaf plane, and the other three rows in front of the leaf.

These geometric and spatial conditions can be used to create even more intense illusions, as shown in Fig. 7.19. This is a modification of Fig. 7.18 by Kitaoka. The differently shaded rectangles are now grouped in whole blocks shifted to each other, which further intensifies this fantastic illusion of movement.

7.9.3 *Rotating Snakes*

The illusion of the rotating snakes is a current figurehead of the new popularity of moving images and comes from the visionary researcher and artist Akiyoshi Kitaoka, who is already familiar to us.

Most of the travelers have seen a picture like the one in Fig. 7.20 somewhere before. The deception works best on a computer screen, but it can also be observed on paper, especially if one looks at Fig. 7.20 closer to the edge of the field of vision with peripheral vision. To make the patterns rotate, you don't have to move the leaf or the head! Just look blurred toward the edge of the picture—after some time you will notice that the circles seem to rotate by themselves.

Fig. 7.20 Rotating snakes. (With the kind permission of Akiyoshi Kitaoka)

There is still no generally accepted explanation for this phenomenon. The term *peripheral drift illusion* (Kitaoka and Ashida 2003) has established itself as an explanatory approach. Kitaoka recognized in his paintings that the sequence of color and brightness of the circle segments is important to make the patterns move with peripheral vision. Thus, the apparent mobility always follows a brightness gradient from black to grey tones and also a brightness gradient from white to grey tones. The circles in Fig. 7.20 move clockwise according to these brightness gradations of the peripheral drift illusion.

The important role of these brightness gradients as a motor for the apparent circular movements leads to the assumption that, as in the Ouchi illusion, spatial replacement mechanisms are involved. Thus, the different brightness levels could stand for different depth layers, to which a different speed could unconsciously be assigned by our perception.

It is worth mentioning that the rotating snakes on the computer screen create a much stronger, almost vivid impression even without peripheral vision. The flickering of the screen seems to really get the snakes rolling. For this reason alone, it is worthwhile to have a look at the very good websites of Akiyoshi Kitaoka.

7.9.4 Whirling Rings

If you are sitting securely, please refer to Fig. 7.21.

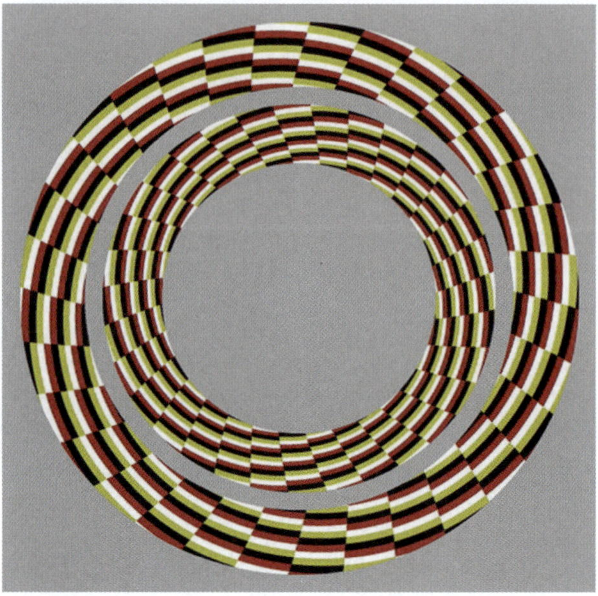

Fig. 7.21 Two whirling rings. (Courtesy of Akiyoshi Kitaoka)

The two ring surfaces begin to move in different directions: The inner ring shrinks and the outer ring expands! Another effect—similar to the Pinna-Brelstaff circles—is visible when you vary your viewing distance. If you move your head toward the blade, the outer circle will move clockwise as if by magic, while the inner circle rotates counterclockwise. When the head movement is reversed, the circles rotate in exactly the opposite direction.

The reason for this fantastic mobility is a mixture of the previous illusions: The peripheral drift caused by the special brightness arrangement, the coloring of the ring segments, and the special orientation provide the pulsation and rotation of the rings.

7.9.5 Heat Flicker

Our journey through motion vision now comes to its end, the bus stops with squealing tires, and the breathless travelers get out.

"To make you really thirsty, I would like to show you a hot picture at the end," says the bus driver. He has an agreement with the drinks salesman, who has already rushed to our bus with a picture in his hand.

All look at Fig. 7.22 together. This picture again connects the most diverse causes of different movement illusions in a clever way: Patterns that repeat themselves again and again are connected with the spatial perceptual specifications (as in

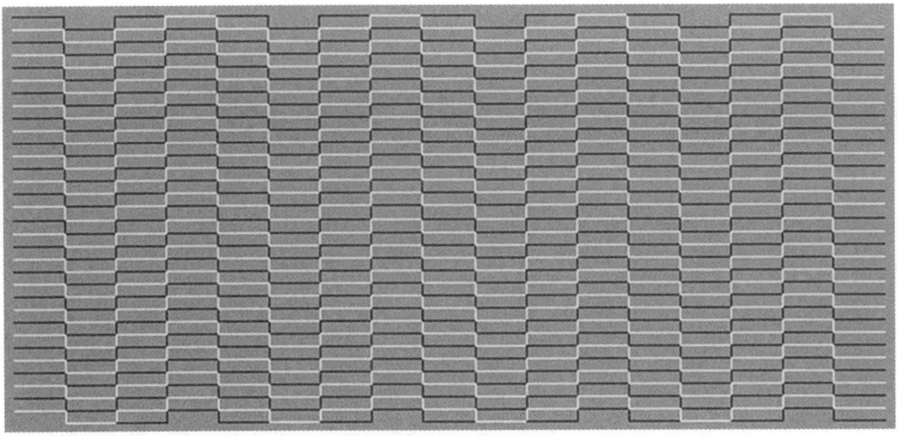

Fig. 7.22 Heat Flicker. (Courtesy of Akiyoshi Kitaoka)

Pinna's pictures) and the peripheral drift illusion (participating colors: black, grey, white). What emerges from this can be seen in Fig. 7.22, which is not wrongly called "Heat Flicker."

Here is the explanation: We recognize the spatial depth of the periodic patterns by the black and white boundary lines and our knowledge of the position of the sun. Thus, the rectangles that have a white upper boundary line are perceived as lying in the foreground. Conversely, the areas that are bounded at the top by a black line are perceived as being deeper.

The concrete cause for the flickering effect is probably the saccadic eye movements—as already known from the pictures with repeating patterns. Due to the persistence of our perception, the resulting different spatial elevations on the retina are temporarily superimposed, covered, or enlarged. This unsatisfactory conflict situation is masked by the impression of flickering/pulsating—which is once again proof of the wonderful inventiveness of our perception. Amplified by the peripheral drift illusion, an intense flickering pulsating effect is created—which looks like the flickering of hot air.

After the tour group has looked at the hot picture extensively, the business with cold drinks and ice cream is going very well, so that all participants are very satisfied with this turbulent trip.

Chapter 8
Eighth Journey: Everyday Life Isn't Grey At All—Illusions in Our Daily Lives

What does everyday life have to do with adventure and discoveries? Everyday life, that's normal life. Office, school, driving—every day the same route, traffic jams, rainy weather, shade and light, hairdresser and dentist, shopping, television. But grey everyday life also consists of many small and big wonders and illusions, if you only look with open eyes. And sometimes our daily life is of course leisure and sport, football stadium or a holiday trip.
Let us take you on an impressive journey through the wonderful world of everyday illusions!

8.1 In the Supermarket

"Everyday life, that's what we have every day—why even consider it in this so far so exciting adventure travel guide?" you might be thinking right now.

Our travel group feels something similar as they meet again (much too) early, but thirsty for action at the bus stop.

"For my everyday life you don't need a coach, at most a public bus," says one of the early risers. But then our fully fueled coach and good-humored bus driver arrive and everyone is sitting quickly and eagerly in their seats. After a few minutes the bus stops again in front of a supermarket. Before the first ones complain if this is going to be a coffee ride, the bus driver quickly walks ahead and into the shop, and we follow.

We are welcomed by a friendly atmosphere with background music, pleasant temperature, and warm light. This is, of course, intended to encourage customers to stay longer. And the longer the customer stays, the more he buys. Normally, the fruit and vegetable counter is located right at the entrance of every upscale supermarket. The presentation of the goods always plays a very special role here. The main thing

T. Ditzinger, *Illusions of Seeing*, https://doi.org/10.1007/978-3-030-63635-7_8

Fig. 8.1 Fruit presentation in the supermarket

is to make the products appear fresh and crisp. Lighting is particularly important in this respect. Of course, customers and our travel group prefer to shop when the lighting and therefore the color of the goods is appropriate. Bananas, for example, should always look as yellow as possible.

Yellow for bananas means fresh and ripe. This is why bananas are often irradiated with yellowish lamps as in Fig. 8.1. This also makes the unripe, greenish bananas appear more yellow than they actually are. However, we as customers are happy to accept this deception in favor of the appealing overall impression. Later, when the banana is outside the yellow light in the shopping cart, you can see exactly how yellow it really is.

"By the way, the shopping trolley can also be used to deceive customers," says the bus driver. "Perhaps you have already noticed the trend toward ever larger shopping carts. Supermarkets have an interest in using the largest possible carts."

"That's logical, so more fits in," says one traveler.

But there is another, almost even more important reason, because very rarely is a shopping cart really full. The purchases appear small and few in a large shopping cart (see Fig. 8.2) and this naturally invites further purchases. Similar to Titchener's illusion in Sect. 2.15, the size of the shopping is perceived differently according to the size of its surroundings, combined with an increase in size contrast.

If the shopping environment is large, the items selected are perceived as small. Conversely, they appear larger in a small shopping cart.

A supermarket works with a variety of other deceptions and tricks that customers are often aware of but are still fooled by again and again. For example, when placing products on a shelf: there are, of course, good and bad places there, as everywhere

Fig. 8.2 Size contrast illusion in the shopping cart

else. The best places are always at eye level and more in the middle of the shelf than at the edge. Experience shows that customers access products from these eye-catching places much more frequently. Of course, the manufacturers know this too and even pay for it—which means that these excellent places are mostly occupied by more expensive branded products. In order to find cheaper products, you should therefore get down on your knees and look at the edge of the shelf. Another customer trick is the arrangement of the product groups in the shop. It has been shown that it is advantageous for sales to first offer healthy goods such as fruit, vegetables, muesli, and others, and to leave candy, chocolate, potato chips, and so on, to the end. These unhealthier purchases are of course made with a much better conscience if you already have a few healthy things in your shopping cart. And of course, one is more likely to buy such small luxury goods if one stands in a long queue at the cash desk and has them in front of the nose. After our travel group has finally survived one of these wanted or unwanted long queues at the cash desk, everybody hurries to get quickly back into the bus to see how the bought bananas look like in the daylight. And everyone hopes that they can somehow make up the time lost at the checkout.

8.2 Save Time

"Here comes the proof that optical illusions can not only be fun, but sometimes they can actually be useful," says the bus driver. He darkens the bus and starts writing on a projector:

> W lv n fntstc, fscntng, wndrfl wrld. ch n f s s n mprtnt pc f th pzzl n ths wrld nd trs t gt alng s wll s pssbl.

"What is this now?" asks our seat neighbor. At first glance, it seems to be an accumulation of incoherent letters. But somehow the whole thing seems strangely familiar again. The first impulse is certainly to proceed analytically and to look at it logically letter by letter. Maybe this "pedestrian method" will actually succeed in making individual words accessible. After some time the group recognizes the general rule that all vowels were simply omitted.

The two sentences are much easier to read, however, if one "steps back a little with one's mind's eye" and simply flies over the text and takes it as a whole into view. You will then probably quickly recognize the two introductory sentences of this book, at least roughly or in larger parts:

> We live in a fantastic, fascinating, wonderful world. Each one of us is an important piece of the puzzle in this world and tries to get along as well as possible.

Our perceptual system shows similar performance in this cognitive challenge as in visual tasks, such as the recognition of the block image in Fig. 3.8, which is a direct visual analogue. Here as there it is about the recognition of the entire context and a good figure (like Marilyn Monroe or even an understandable text), whose carrier in Fig. 3.8 is the individual color areas and here, the individual letters, whose meaning only becomes clear when viewed with a little more "distance."

This astonishing cognitive performance of our perceptual system enables us to recognize texts that are written in a very incomplete way. In our example, only 82 of the 134 letters of the original text were used, which is a 39% reduction in the amount of information that would ideally have to be written! Of course, the time added to the writing is at the expense of the time lost in reading.

In the meantime, the bus driver has again typed something new at high speed:

> "To thrs end, xe use tur sezses, wqich eoable ys to skmultcneoutly pejceivm an enermousy amoupt of ekvirobmentzl stiruli asd infjrmapion."

Recognizing this text is somewhat more difficult for the passengers than in the first example. This time the text has wrong letters instead of simple letter gaps. This is exactly the effect that occurs when you type (too) fast—the faster, the more incorrect. And here, every fifth letter in the text is wrong. Set right, it says:

> To this end, we use our senses, which enable us to simultaneously perceive an enormous amount of environmental stimuli and information.

Now the bus driver increases his writing speed and thus the error rate and shows us now:

Thx hufan rysiem hf pvrcsptjon was levdloded en tpe ciurwe og evplurioq in kloce icterxc-tqon aitp thep enpirpnmxnt.

That's really a tough nut to crack, with a typo on every third letter. With this we have certainly reached our perceptual limit, but probably also the meaningful spelling error limit. Again, the bus driver made it easy for himself and simply took out the next sentences of the preface of this book. This is the sentence that coincides with the driver's mistake-ridden sentence.

The human system of perception has developed in the course of evolution in close interaction with the environment.

Our perception has also proven to be very effective in recognizing meanings and sentences and can easily compensate for a large number of letter gaps and typos. This remarkable feature actually leaves some leeway to gain time in our ever faster media landscape. For example, there is a growing tendency among frequent writers in the working world, who have to read and write several hundred emails a day, to consciously accept spelling mistakes. As long as the content is sufficient and time is saved, bad form and small mistakes do not disturb their process.

The situation is completely different at our next travel station, where nobody wants to experience major mistakes.

8.3 At the Dentist

"Man, we are really not spared anything," cries one of the travelers as we all take a seat in the waiting room of a dentist.

"What do I have to say then, I'm here all day," replies a man in a white dentist's coat who just walks in the door. "After all, I have made an effort to design my practice according to the latest perceptive psychological findings!"

The room is mainly in blue and green, and in fact this has a calming effect on the travel group.

8.3.1 Room Colors

Probably there are not many rooms whose design is perceived with as much awareness and time as the waiting room and the treatment room of a dental practice. In addition to the pure functionality of a room, the emotional aura and coloring of a room should therefore also be taken into account. A first clue in this direction is given by surveys on the favorite colors which was done for Germany. Various studies have repeatedly shown a very clear vote for blue. In a study by Heller (2004), the result was 38% blue, 20% red, 12% green, 8% black. The least popular colors were brown (29%), followed by orange and violet (11% each). In a first approximation, patients should therefore feel most comfortable in a room with mainly blue and some red, green, and black. Of course, one should always consider the expectations of the visitors to the respective room. But what exactly do visitors expect from a room? How can this be translated into colors? Again, quantitative psychological surveys are helpful in answering these questions.

In the aforementioned study by Heller, the test subjects were asked about their color associations with given terms. For example, the color blue was assigned to the term "sympathy" by 28% of the test subjects, red by 17%, green by 16%, pink 9%, white by 9%, and violet by 8%. A total of 200 terms were assigned colors in this way. Depending on the patients' expectations of the respective room, the results of the study can be used to achieve a very fine recommendation for color design.

The expectations of a room can be achieved, for example, from a patient survey. For a waiting room, for example, one can imagine that the expectations of the room could be: "calming effect, friendliness, helpfulness, hope, optimism, cleanliness, safety, sympathy, trust and reliability." If one adds the respective color assignments of these properties from Heller's study, the color weighting for the expectation of a waiting room is as follows (see Fig. 8.3 upper row): 20.9% green; 20.4% blue; 18.0% white; 5.6% yellow; 5.2% red. This means that, for our example, a waiting room should be primarily in the colors green, blue, and white.

In the treatment room, of course, the focus is on pure functionality. To be able to perceive and detect fine structures of the teeth, there should be no interfering colors in the vicinity of the dentist's field of vision. The neutral colors white and grey and the natural background colors blue and green are recommended. Signal colors such as yellow, red, or violet should be avoided. The argument in favor of using the background colors blue and green is that the complementary colors to the tooth colors lie in the uncolored blue-green area. This results in a strengthening of the edge contrast enhancement and an accentuation of the tooth colors, and thus a sensitive possibility of perception of the smallest color deviations in this area.

But the emotional color perception of the patients plays a role here as well. Because even in the treatment room the patient usually has enough time to perceive color moods accurately. The travel group selects ten typical characteristics for this example again. The choice is: "Honesty, functionality, accuracy, concentration, performance, safety, cleanliness, trust, science, reliability." The result for this example combination is: white 26.3%, blue 19.8%, green 8.7%, black 6.5%, gold 5.1%, grey 5.1%, silver 5.0% (see Fig. 8.3 below). This means that for our example, the treatment room should be kept mainly in white and then in blue. Green no longer plays the same role as in the waiting room, but can be used as an embellishment alongside

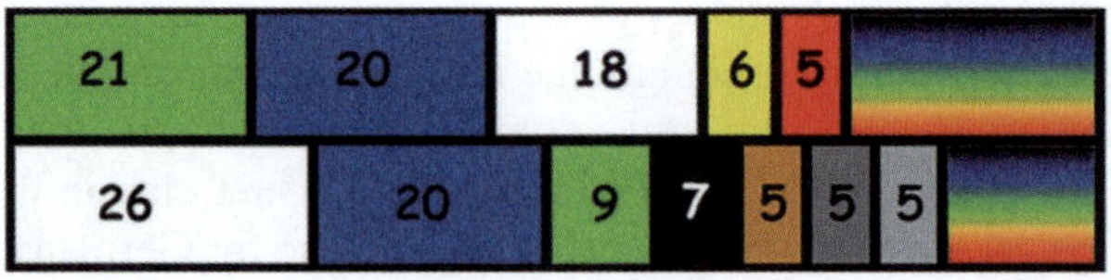

Fig. 8.3 Typical color expectations of a waiting room (upper row) and a treatment room (lower row). The thickness of the colored bars corresponds to the respective percentages. Colors with percentages below 5% are not listed individually but are included together in the area with rainbow colors

black, gold, grey, or silver. Remarkably, this emotional assessment corresponds very well with the functional specifications. Our color subconscious seems to have a good portion of color consciousness here!

The dentist is now really getting into the swing of things because he is leading into his special subject, the determination of tooth colors. He even forgets his next patient, but at least he is waiting in perfectly colored rooms.

8.3.2 *Tooth Colors*

"Determining the right tooth color was still a challenge, even with my ten-thousandth crown," the dentist exaggerates a bit. And this despite the fact that the range of tooth shades is actually a very limited size with only very small shades: it has been shown that humans—and even dentists—are only just able to distinguish 100 different tooth shades by direct comparison.

The schematic extent and size of the tooth color space can be seen in Fig. 8.4. The sphere shown there contains the natural color space of all colors visible to humans. The application method is based on the international color standard CIE Lab [CIE No. 13–15, Commision Internationale de l'Eclairage, Bureau Central de la CIE, Paris 1971] and is carried out on the basis of the three variables: brightness, color intensity, and hue. The brightness is applied from the bottom (dark) to the top (light). The intensity is plotted from the center of the sphere to the outside, with the degree of intensity increasing from the inside (intensity 0) to the outside. Finally, the hue—the pure color tone—is applied in a circle around the axis of the brightness sphere.

Fig. 8.4 Total color space and tooth color space represented in the three dimensions brightness (from bottom to top), saturation (from inside to outside), and hue (circular around the sphere)

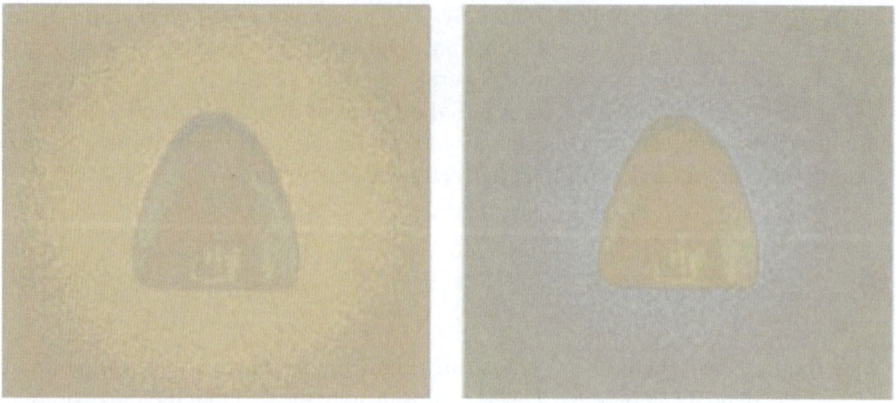

Fig. 8.5 One bluish and one reddish tooth, right?

If you look very closely into the bright uncolored red-yellow area of the sphere in Fig. 8.4, you can see a small banana-shaped figure. Exactly within this narrow banana skin lie the 100 discernible tooth colors. Despite this clear number and extension of the tooth color space, you will immediately see that fantastic illusions are also possible here, which are sometimes even more pronounced than in the overall color space due to their fine nuances. In the following pictures, the dentist shows us some examples, which all originate from the tooth color space.

"First look at the typical situation illustrated in Fig. 8.5 for determining the tooth shade," he says.

Assume that the left tooth is in the patient's mouth and the right tooth is used for color determination with slightly different backlighting. In this case, two teeth of different color and brightness are clearly visible. However, this clear, intuitive perception is an illusion, because the two teeth are exactly identical! If you want to see for yourself, cut out a mask from a white paper with scissors, leaving the two teeth free, and hold it over the picture!

Everyone is amazed at the extent of this illusion, which is almost stronger than the simultaneous color contrast in the entire color space, which we had observed in the fifth voyage (Fig. 5.28), or the brightness contrast enhancement of the two suns in Fig. 3.22. "First of all, you can see what kind of difficulties we have to deal with every day," the dentist picks up the ball.

And there are many more difficulties.

"Next, please consider the teeth shown in Fig. 8.6 against a background which, could have been caused by solar radiation from above. The price question is: 'Which of these teeth appears to you to be uniformly colored from top to bottom?'" asks the dentist.

You are probably betting on tooth B. The true conditions can be shown simply by fading out the background. See Fig. 8.7, in which a monochrome mask was placed over the background of Fig. 8.6. It is now clearly visible that tooth A has a uniform

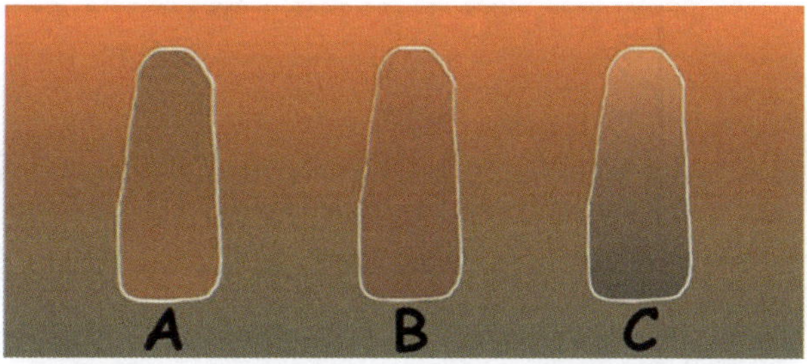

Fig. 8.6 Which tooth is evenly stained from top to bottom?

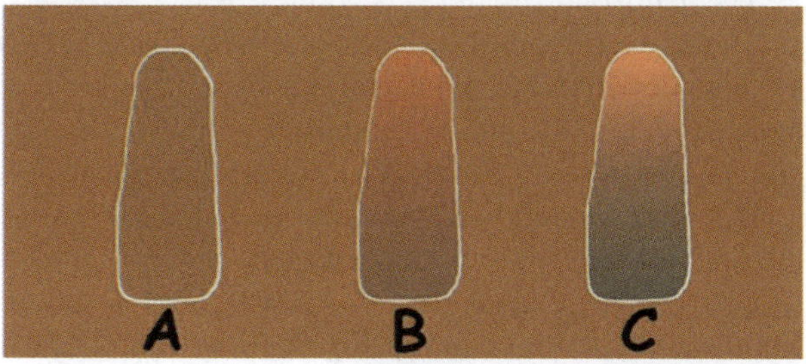

Fig. 8.7 The resolution with uniform background: tooth A

Fig. 8.8 The left tooth appears darker and redder than the right tooth. In reality, however, the two surfaces are colored identically except at the transition edge. You can convince yourself of this by covering the edge with your index finger or a pencil

color. As you can already see: even our perception of tooth color sometimes has little to do with reality!

The influence of unusual lighting conditions on color perception can be seen in Fig. 8.8. It shows two differently illuminated surfaces (e.g., two adjacent teeth), which can be clearly distinguished from each other. The left tooth appears darker and redder than the right tooth.

In reality, however, it is a deception again. To see for yourself, simply cover the vertical transition area with your finger or a pencil. You can immediately see that the two surfaces are absolutely identical in color and brightness. The surfaces are illuminated differently only at their dividing line. We already know this deception from Journey 3 as Craik-Cornsweet-O'Brien deception. Here, too, the effect is not stronger in the general shade space than in our fine and sensitive area of the tooth shade space.

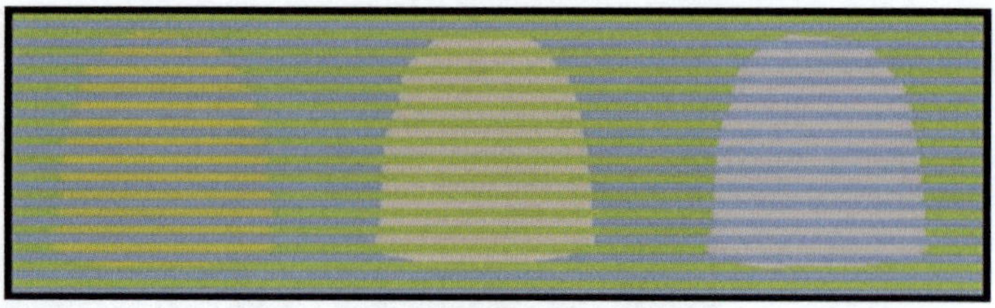

Fig. 8.9 Which tooth corresponds to the middle, the left tooth or the right tooth?

The dentist now really gets going.

The next example shows three teeth against a background that could be created by a window blind in the dental practice. The question is: Which tooth in Fig. 8.9 looks similar to the one in the middle—the left one, or the right one? You will certainly notice a slight difference in shade between the two teeth.

However, the left tooth seems to be much more similar to the middle tooth than the right one, which looks almost blue.

But also this time we are wrong with our appearance! In reality, the right tooth has the same tooth color! In Fig. 8.10 we simply pull up the roller blind slowly. After just a few centimeters it becomes clear that the right and not the left tooth corresponds to the middle tooth.

Everyone is now deeply impressed by the extent of the dental deceptions presented and the performance of the dentist to somehow overcome these problems every day. After all, he tells us a trick that helps him to get along better and ultimately even provides an explanation for the intensity of the deceptions in the tooth color space.

He says, "You have to know that our natural color vision of any object normally works by first determining the pure hue, then the brightness and then the intensity/luminosity. This procedure seems to us to be the natural, intuitively correct one. The color shade is always in the foreground. For example, there are many names and designations for the hues, but none for the brightness (except perhaps black, grey, and white) or intensity. For the tasks of tooth color determination, however, it has been shown that this purely intuitive human color perception is extremely error-prone, especially in rooms with artificial lighting."

The reason for this lies in the special shape of the tooth color space, which is similar to a high-standing banana in the bright, uncolored, reddish-yellow area. If you take a very close look at Fig. 8.4, you will see that this is actually a flattened banana with a strong squeeze in the hue range. Due to this strong limitation of the shade, it is understandable that our very successful recipe for categorization by shade for the entire shade space is not recommended for determining the tooth shade.

Rather, in the first determination stage the size with the largest span should always be determined, then the size with the next largest variance is considered, and so on.

For this reason, the sequence is recommended for a structured tooth shade determination:

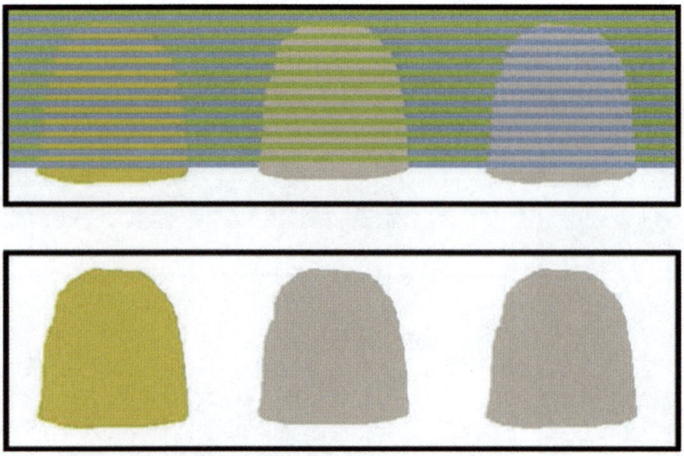

Fig. 8.10 The roller blind is pulled up and after a few centimeters it becomes clear that the right tooth corresponds to the middle one

(1) Brightness, (2) Saturation, and (only at the end) (3) Hue.

The bus driver says: "This is similar to getting into a bus, or even better into a train, which has a similarly long form like the tooth banana has. In order to find a reserved seat quickly, we always look for the car number first and thus limit the train length considerably. Only in the next steps do we look for the right compartment and then the seat." Although this method of preferentially determining brightness strongly contradicts our natural color perception, it offers decisive advantages in tooth color determination with a significant reduction in the error rate. This can also be demonstrated in all of the examples we have presented.

For this purpose, we consider the pure brightness of the previously presented tooth illusions (see Fig. 8.11), which can also be approximated by black-and-white photograph or photocopy of the images. It is clearly visible that the teeth are now identical and that the illusions observed earlier are thus dissolved.

Before his visitors dissolve in the direction of the bus, the dentist clears his throat and, as befits an exciting adventure trip, announces a few highlights for the finale.

"Please look at the two opposing teeth shown in Fig. 8.12. Surely you can see a clear difference in the coloring of the two teeth. The upper tooth looks clearly darker than the lower tooth—but it is not. If you hold your index finger horizontally over the parting edge between the two teeth, you can simply make sure that the two teeth are exactly the same brightness," he says.

This fantastic illusion is created by a clever combination of the Craik-Cornsweet-O'Brien effect with the impression of illumination from above (visible in the background and on the teeth, including shadows).

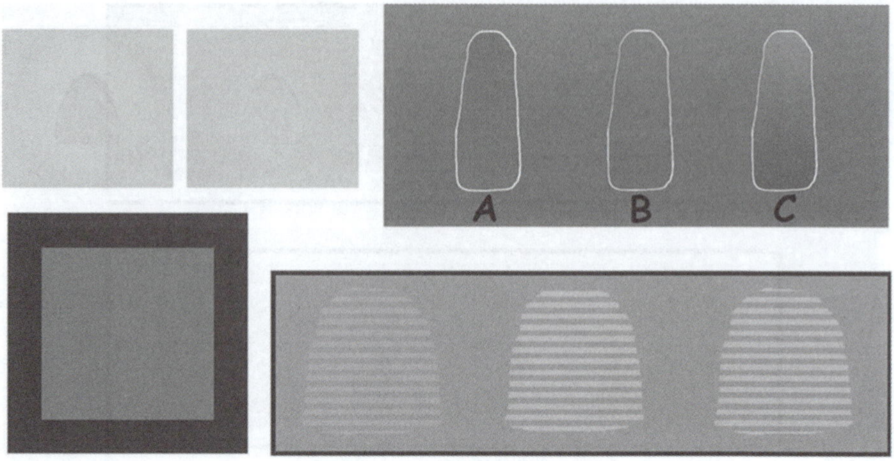

Fig. 8.11 The previous deceptions in black and white—the deceptions simply disappear!

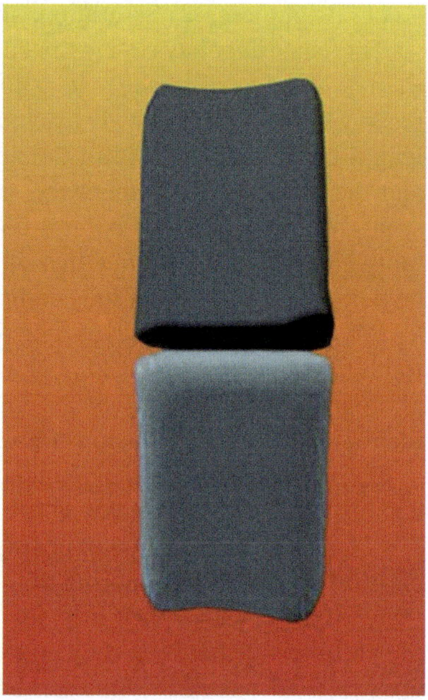

Fig. 8.12 One dark and one bright tooth—right? For resolution, please just hold the index finger over the contact edge between the teeth! (Based on a template by Beau Lotto, with friendly permission)

The illusion in Fig. 8.13 works in a very similar way. In the picture above, two darker teeth are clearly visible, which should actually be extracted soon. The individual teeth also look somewhat inharmoniously arranged. By having a closer look we recognize the reason: the illumination is not uniform, some teeth seem to be illuminated from the left and some from the right. By fading out the lighting effects,

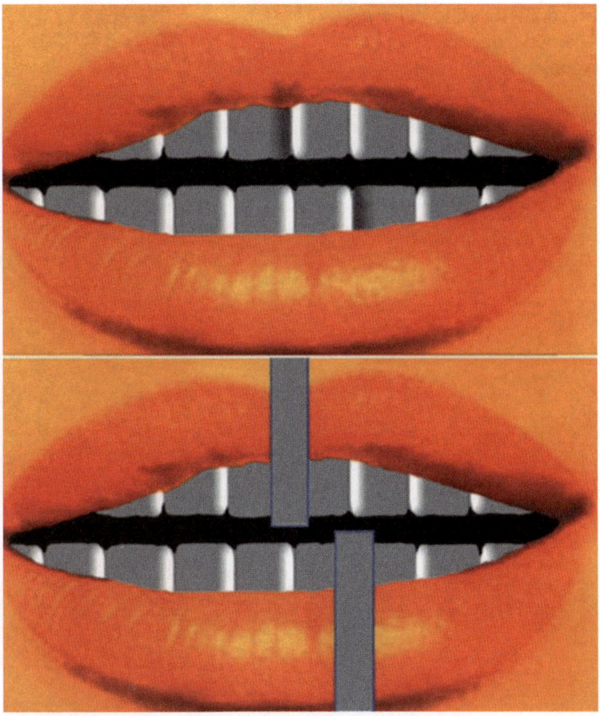

Fig. 8.13 Top: two dark teeth, hopefully not to be pulled. Below: fortunately it was only a deception

as it is the case in the lower partial image, one immediately recognizes that it is again one and the same tooth color. Thus, the teeth can stay in the mouth!

Our dentist notices that he has a waiting room full of patients who are slowly getting impatient. Even a nice waiting room and all the theory doesn't help anymore. He has to go back to work and leaves us with some further reading (Ditzinger 2004a–d) and one last fantastic picture, Fig. 8.14. At the top there is clearly a dark, reddish tooth, and at the bottom a light, yellowish tooth. But the two teeth are completely identical! The explanation for this astonishing illusion is, on the one hand, a strong color gradient in the background from yellow to red as with a simple illumination from above. In addition, the two teeth have an inverted color gradient from red to yellow, but only with half the span, which has turned out to be a kind of optimum for achieving a maximum deception effect. The appropriate combination of the two color gradients creates a strong interplay of edge contrast enhancement and the impression of illumination from above, which makes the lower tooth appear as directly illuminated, standing farther forward and appearing much brighter than the upper one, which seems to be in the background and in dark shadow.

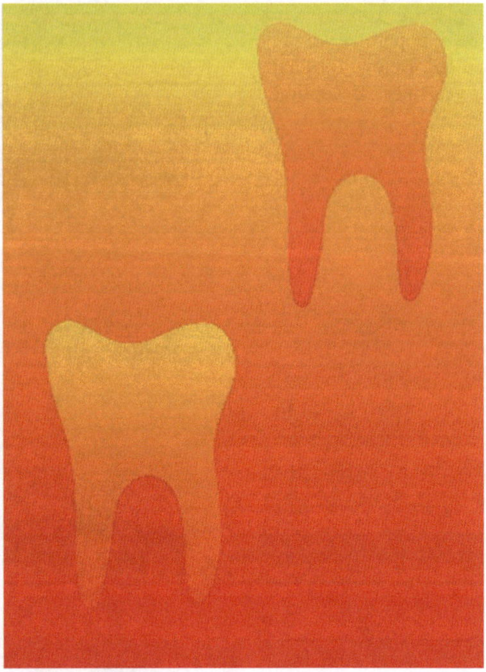

Fig. 8.14 Hard to believe: The two teeth are completely identical!

8.4 In the Stadium

Because of all the tooth colors, our bus driver almost forgot that his favorite team wants to pull the tooth of its opponent today. Quickly everyone runs into the bus and he now accelerates powerfully. Now it's getting really serious and important—it's all about football! Week after week, football captivates millions of people. And this not least because of its spectacular deceptions! The attacker deceives the opponent again and again with new clever body deceptions and the referee with dives or God's hand. The simplicity of the rules of the game allows severe mistakes to be made with a fixed regularity, which, as with the Wembley goal of 1966, become real legends. Especially in matches between Germany and England, the goal lines, in the referee's eye, seem to bend by meters every now and then.

8.4.1 The Round Must Go into the Square

But often the round just doesn't want to get into the goal. An interesting view can be seen in Fig. 8.15 on the left. So you can imagine the situation of a shooter at the penalty spot. The goal becomes smaller and smaller and not even the ball appears round. In the picture, the actually perfectly round ball shape (see Fig. 8.15 at the

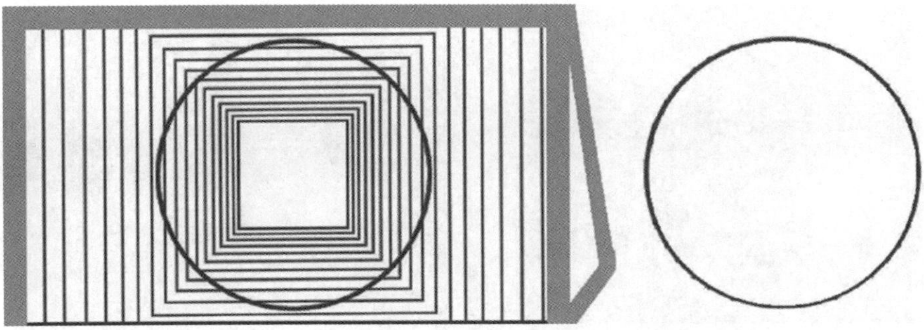

Fig. 8.15 Left: The round one has to go into the square—although sometimes it does not seem so easy. The ball looks strangely angular, although it is exactly circular; see right

right) is perceived as clearly angularly distorted by the net lines of the background, and the ball almost looks like a diamond standing on edge.

The reason for this intense deception is, of course—apart from the shooter's fear of the penalty—an optical illusion. Similar to the Hering deception described in Sect. 2.7, the arrangement of lines in bundles of the background interact with the foreground in our perception. Again, our perceptual system strives for the best possible Gestalt with ideally right angles of intersection between ball and background. In order to achieve this optimum as well as possible, it simply changes the perception of orientation again and tilts the position of the lines at all intersections—as good as possible for the overall impression.

This apparent tilting process takes place simultaneously at all intersections. This creates the overall impression of a "square" ball standing on a point.

"That's all no problem—to get the ball into the goal, the shooter just has to work with spin," says the bus driver, and you could think that he has already played with the players in the German Bundesliga himself, who are just about to enter the packed stadium. The tour participants make it just in time for the kickoff.

8.4.2 Cam Carpets: The Camera Carpets

Unfortunately, the game is developing disappointingly so far and once again the ball misses the goal by several meters. A player jumps after him quickly and seems to crash head-on into the advertising board behind the goal line as shown in Fig. 8.16. But nothing happens—the player just seems to dive through the board and a second player just stays on it.

The bus driver already explains: "These advertising boards as in Fig. 8.16 for 'TelDaFax' and 'diedruckerei.de' are in reality carpets lying flat on the floor. If you look at them from a certain angle—in which the main stadium camera is positioned—they appear to stand three-dimensionally like a real advertising board. These carpets are therefore called Cam Carpets. This fascinating spatial illusion is

Fig. 8.16 Two CamCarpets ("TelDaFax" and "druckerei.de"). (Courtesy of Bayer 04 Leverkusen Fußball GmbH)

based on a skillful application of the theorem of rays, which is calculated exactly to the camera position. The carpets are particularly popular in advertising as they are located in the immediate vicinity of the goal. The cameras show this area particularly often—which of course also means that the advertising banner is often in the picture. For the clubs and above all the players, the significantly reduced risk of injury is also an advantage compared to a real advertising gang standing in the way."

For stadium visitors outside the camera area, the illusion quickly dissolves, and one clearly sees a flat and strangely distorted carpet lying there.

The bus driver is once again in top form today and has brought us a paper with the lettering "Illusion"; for an explanation, see Fig. 8.17.

"We now want to spatially erect this advertising lettering for our book," he says. All we have to do is rotate the lettering slightly and then increase its height. Soon you get something like in Fig. 8.18. The camera carpets look similarly distorted when you look at them from above or away from the camera.

Please take a look at this camera carpet from the bottom left—this is roughly the perspective of the main camera—and you will see that the whole thing spatially arranges itself. This impression is further enhanced by the grey auxiliary surface, which is intended to provide a spatial foothold. You can also find similar foothold areas at the bottom right of the "TelDaFax" and "diedruckerei.de" in Fig. 8.17.

The game is otherwise a disappointment and so we are happy when the bus driver announces our departure in order to track down more illusions happening in our leisure time.

Fig. 8.17 "Illusions." This lettering is to be represented spatially

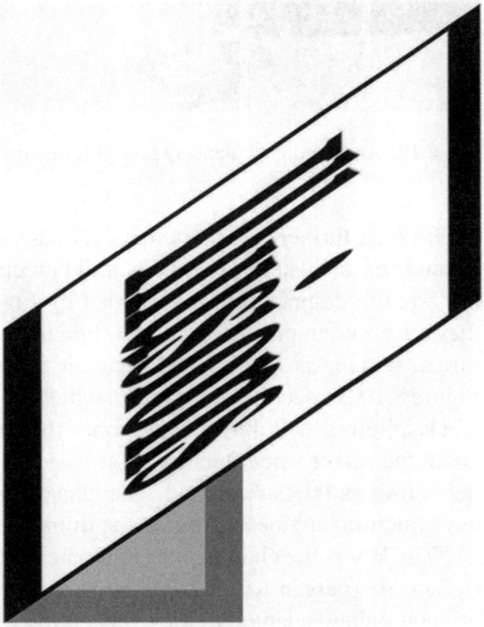

Fig. 8.18 The camera carpet for Fig. 8.17, created from a simple rotation and vertical stretching

8.5 Spatial Misinterpretations During Vacations Time: In San Francisco and Skiing

We just took a seat in the bus again, when the bus driver says: "I just came back from San Francisco and from skiing." He proudly shows us some holiday pictures. Of course we are all wondering how a bus driver can afford all this, but already the pictures cast a spell over us. Figure 8.19 actually shows cars with California license plates. In front of them stands a person who seems to have been copied into the picture at a completely impossible angle. But the bus driver assures us that it is a real picture. As the cars, the shadows and also the bushes in the background provide us with a clear spatial perception, we accept the physically actually impossible inclination of the person.

Fig. 8.19 An incredible balancing act—this picture is not a photomontage!

We hide further inconsistencies like the clearly sloping house wall in the background for the benefit of good overall perception. The proof that Fig. 8.19 is in fact not a photo composition is provided by Fig. 8.20. It shows the real conditions as they are photographed every day by hundreds of tourists in San Francisco's Lombard Street, which, as in Fig. 8.20, has an extreme gradient of about 15° and more. Figure 8.19 is a section of Fig. 8.20, which was then simply rotated clockwise by 15°.

Despite our new knowledge about the extreme road gradient, we also wonder about the driver's next picture (Fig. 8.21). This time everything is straight—except the houses in the background. They have a gradient of 15°, which is incredible for any structural engineer! After some thinking, a travel companion gets the solution:

"The key is the changed orientation of the person in the foreground!" he shouts. In fact, the person in the foreground for this photo has significantly changed his orientation and adapted to the cars and the road. If you rotate the picture clockwise by 15°, as in Fig. 8.19, you get the impression that the houses in the background must be diagonal—although in reality they were the only vertical in the picture.

As you can see, the perception of spatial orientation is strongly linked to our experience and therefore prone to error in extreme situations. But even with less extreme inclinations, we repeatedly experience spatial orientation deceptions in our environment. The bus driver shows us a picture from his last skiing holiday (Fig. 8.22). The picture shows a well-traveled ski slope with a slight gradient—but in which direction? Even after longer viewing, the situation is not clear. The spatial clues are simply missing.

The skiers in the foreground ride and stand in different directions. So the skier in the front seems to ski backward, while the second and third skiers seem to ride forward. The skiers farther in the background, on the other hand, cannot be resolved clearly. As already seen in Journey 4 on ambiguous perceptions, objects at a distance or without a clear surface texture can be perceived both as being turned toward or away from us (see Sect. 4.4). For example, an airplane in the air at a great

Fig. 8.20 The overall picture in correct horizontal orientation, as taken daily by tourists in San Francisco. The road here has an extreme inclination of about 15°. Figure 8.19 is simply a section of the image and was rotated 15° clockwise

Fig. 8.21 Oblique houses. The difference between this image and Fig. 8.20 is above all the orientation of the person who has adapted to the road and the cars. If you now turn the original picture clockwise by 15°, the houses appear clearly oblique—although in reality they are the only vertical in the picture

Fig. 8.22 In which direction does it go downhill? The skiers do not provide any clear directions of movement and there are no further spatial clues from the sun, light, or shadows. A decision is not clearly possible on the basis of the picture. Only the posture of the third skier could make a slight difference for a slope toward the photographer

distance can be perceived either as flying toward us or away from us. And the same is now the case with skiers who are farther away. The decision is further complicated by the absence of the sun and the resulting casting of shadows. The extremely important role of the interplay of light and shadow will be further illustrated in the next pictures.

8.6 Sun, Light, and Shadow

In the sixth journey we saw light and shadow play a big role in our recognition of depth (see Sect. 6.5.5).

There we have already seen that the knowledge of the position of the sun is a powerful element of our perception. Incoming information is always interpreted by our perceptual system in such a way that the sun appears to be at the top and the bright surfaces appear to be facing the sun. But that is not all.

Another powerful secret of success of our everyday perception is the recognition and interpretation of the shadows cast.

Please have a look at Fig. 8.23 on the left, by the American perception researcher Edward Adelson. On it you see a green cylinder standing on a checkerboard. The cylinder appears illuminated from the top right and casts a corresponding shadow over the small chessboard.

Fig. 8.23 The Checker-Shadow Illusion by Edward Adelson, first published online 1995. (Courtesy of Edward Adelson)

"So where is the deception now?" we ask ourselves. In fact, we have a consistent and coherent perception of alternating white and black chessboard squares in front of us. For example, square A is clearly recognized as dark and square B is clearly recognized as light.

But the physical reality looks completely different! If you cover up the shadow cast as shown in Fig. 8.23 on the right, you can immediately see that surprisingly field A and field B have exactly the same brightness!

How is that possible? Quite simply because our perception is not just a photometer that accurately measures and stores brightness and color tones, but much more. Our brain has the wonderful ability to cognitively evaluate and interpret the conditions of shadow-casting. Over millions of years, this property has proven to be essential for survival, as it was the only way to reveal hunters hidden in the shadows in time. As soon as our visual system has recognized the shadows cast in Fig. 8.23, the fields are reinterpreted internally to arrive at a conclusive perception. Thus, the presumed illumination outside of the shadow cast (as in field A) is simply subtracted internally in order to arrive at an assessment normalized by our perception and independent of the apparent illumination. As a result, the two identical fields, A and B, now appear clearly different.

Just how powerful our internal processes are in dealing with the casting of shadows can be seen in the fantastic image in Fig. 8.24, which comes from the London perception researcher Beau Lotto.

Please have a look at the big magic cube.

Again you can see an illumination (from the top right) with a corresponding shadow cast, which makes the whole front of the cube appear in the shadow. Again, our perception corrects the light conditions according to its successful knowledge from experience. Thus, the green, red, yellow, and blue mosaic elements in the foreground appear in harmony with the directly illuminated elements above. The only exception is the central element in the middle, which appears brown at the top and bright orange in the middle—much brighter than the five yellow elements on the front!

Fig. 8.24 The Lotto Cube. Look at the two middle mosaic pieces of the two visible cube faces. The top mosaic piece appears brown and the front orange—but both are in fact identical. (Courtesy of Beau Lotto)

Fig. 8.25 The proof. (Courtesy of Beau Lotto)

In reality, however, both elements are colored exactly the same. You can convince yourself of this in Fig. 8.25, in which the other elements and the shadow are completely hidden. In fact, both elements now appear in a uniform orange color. By this fascinating cube picture you recognize that the influence of light and shadow-casting extends far beyond the perception of pure brightness to—very impressively—the perception of color.

"Brightness, color, and shadows also play a role in our next station of illusions in everyday life. In particular, it's all about looking the way you want to look with little effort and with the help of some already familiar illusions," the bus driver says, announcing the next travel stop.

Some of the passengers are listening attentively and hope that their problem areas will soon be freed—if not for real, then at least by some kind of beautiful illusion.

8.7 Optical Illusions in the Fashion Industry

Fashion designers like to use optical illusions to make their customers look more individual or better. With simple tricks and illusions, pregnancies can be concealed for longer, those who think they are too slim can be made to look thicker, and those who feel they are too large can be made to appear slimmer. A very simple and popular trick is the implied waist, which immediately makes the wearer look significantly slimmer (Fig. 8.26). Another popular deception used in fashion is the Müller-Lyer deception, which we have already encountered in detail in the second journey (Sects. 2.2 and 2.4): Lines that end with an acute angle (arrow) appear shorter to the observer than lines of the same length with a blunt V-end.

This is why a shirt with a V-neck, for example, makes its wearer appear much longer (Fig. 8.27).

A third simple trick is based on the Oppel-Kundt deception. We hinted on its effect on fashion in Sects. 2.12 and 2.13: The distance between two boundary lines appears much longer if it is filled with parallel intermediate lines as shown in Fig. 8.28 left. Therefore, uniformly colored clothing, as in Fig. 8.28 right, makes the wearer appear slimmer than clothing with vertical stripes!

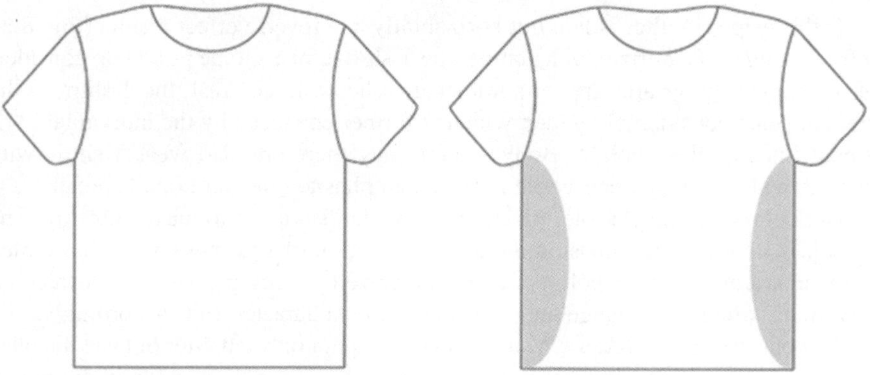

Fig. 8.26 Waisted T-shirts make you slim

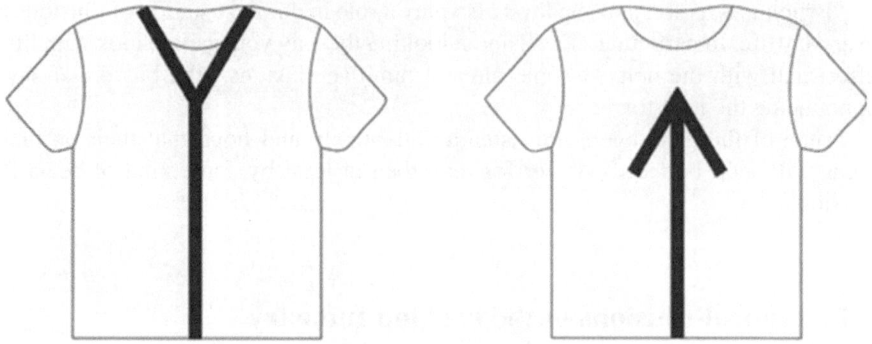

Fig. 8.27 The left T-shirt with V-neck appears clearly longer than the right one with the arrow. The reason for this effect is the well-known Müller-Lyer deception

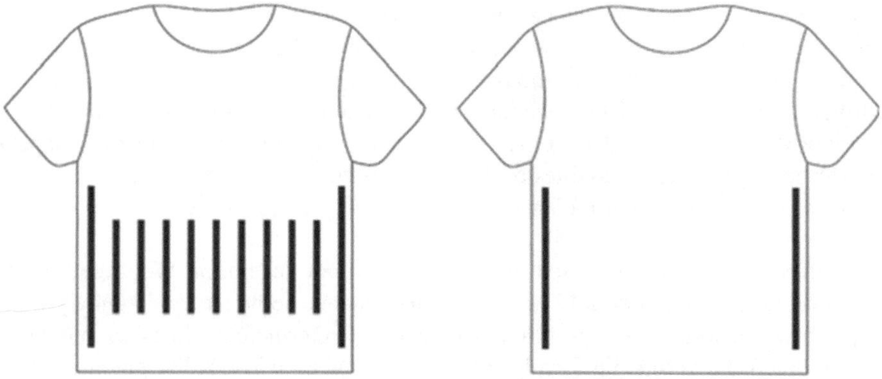

Fig. 8.28 Vertical stripes (left) make wide in comparison to a uniformly colored T-shirt (right). The reason for this deception is the Oppel-Kundt deception

If the striped coulter is laid out horizontally, the reverse effect results (Fig. 8.29 left). Owing to the horizontal hatching, the T-shirt appears to be pulled up considerably, making the wearer appear slimmer than he is. In contrast, the T-shirt on the right appears considerably wider, which is further enhanced by the individual horizontal lines at chest height. Bodybuilders may therefore also wear T-shirts with simple, wide writing at chest height, which emphasizes the horizontal spread.

But colors also play a role. Black, in particular, is known to make you slim. The reason is its low light emission and the elimination of shadows cast by unwanted body curvatures! Other colors can be perceived as being spatially elevated or "warmer," which can change the physique or even character traits accordingly.

The bus driver concludes: "You see, fashion is not only a matter of taste, but also a field of verifiable psychology of perception—a small consolation for all those with 'bad' taste. Besides, most of the time the taste is not so bad, but simply different!"

Fig. 8.29 Left: Horizontal stripes make the T-shirt look slim and elongated. Right: The T-shirt appears much wider and shorter. The cause again is the Oppel-Kundt deception

With that, he accelerates and brings us on the way to a completely new and impressive perception effect.

8.8 The Perspective of Double Images

"The perspective deception that I'm about to show you is actually not a deception at all. Rather, it is a key experiment for understanding our spatial perception," the bus driver says.

The idea for this experiment goes back to Kingdom, Yoonessi, and Gheorgiu, who won the "Illusion of the Year Contest" with a picture of the Leaning Tower of Pisa in 2007. They have simply placed an identical copy of the image next to the original. If one looks at this double image, the second inclined tower appears clearly more inclined than the first.

You can quickly understand the mechanism of this effect by looking at Figs. 8.30 and 8.31. Both pictures simply consist of a source image and its copy.

Figure 8.30 shows a waterslide in an adventure park, which in the left partial picture seems to come from the left and in the right partial picture from the middle.

Figure 8.31 shows a boy running on a footpath, which seems to lead to the left in the left part and to the right in the right part, although this is also the same picture of the path! So it's not two different shots, it's the same shot twice! Our perception, however, always recognizes and treats the two photographs as one overall spatial scene, drawing on its wealth of experience in the perception of perspective. Thus we know that two parallel three-dimensional objects, such as two high-rise towers or lamp posts standing next to each other, have in their two-dimensional representation in the distance converging lines of alignment. The images of two waterslides or paths that are parallel in the three-dimensional reality would therefore be inclined

Fig. 8.30 The double waterslide: one slide seems to come from the left, the other from the middle. In reality, these are two identical images placed next to each other

Fig. 8.31 The two ways. One path seems to lead to the left, the other to the right. In reality they are two identical images of one and the same path!

toward each other. But this is of course not the case with our identical double images under consideration.

The depicted objects are merely parallel in their two-dimensional representation, which our perception recognizes as a spatially diverging tilting in 3D. These images thus show less a deception than the constant striving of our perception for the good Gestalt, and thus the perception of the two individual images as a whole scene. Our wealth of experience in perspective perception then inevitably leads to an apparent tilting apart of the two individual elements.

This effect also works upright. An impressive example can be seen in Fig. 8.32a using a photograph of the Pauliner Church in Göttingen, Germany. The image below was simply added to the image above. The shelves in the upper part of the picture appear to point upward and backward and the shelves in the lower part of the picture appear to point downward. The explanation is the same as before: The two individual pictures are again perceived as a spatial overall scene. And from our perception of perspective we know that the lines of alignment of spatially parallel objects converge on the paper—like the individual lines of the shelves in the lower part of the image, for example. Parallel lines of alignment, on the other hand, such as those created by the juxtaposition of the two images, create the impression of spatial perspectives shifted vertically apart, as can be seen in Fig. 8.32a.

The spatial tilting effect is significantly enhanced if the seating and the floor are omitted in the upper picture (see Fig. 8.32b).

Fig. 8.32 (**a**) The double Pauliner church. Two identical image sections are arranged one above the other. This makes the bookshelves appear more tilted upward in the upper drawing file and more tilted downward in the lower drawing file. (**b**) The same procedure for two identical images lying on top of each other. However, in the upper picture the seating and the floor were cut out. This significantly increases the spatial tilting effect of the shelves. The original photo was taken by Martin Liebetruth, SUB Göttingen

Fig. 8.32 (continued)

8.9 Hiding and Camouflage

After an adventure park, walking races, and church visits, the participants are visibly exhausted and do not notice any of the hidden animals along the way on their way back.

However, this is no wonder with some animals, because many animals are true camouflage artists (Fig. 8.33). With their body coloring and texture, they are perfectly adapted to their natural environment and thus naturally have an evolutionary advantage and always some protection from their hunters.

There are also hidden images in art, such as in the wonderful Fig. 8.34 by Nicholas Wade, who in an ingenious and artistic way combined the effect and the explorer of a deception (Wade 2011).

On the left side of Fig. 8.34 you can see a colored version of the Hermann grid with yellow lines and blue background. Surely you will quickly see the virtual blue points on the crossing points of the yellow lines. We have already encountered these flickering points in the third journey (Sect. 3.3.6). If you take a closer look at the yellow lines, you will recognize the extraordinary peculiarity of this picture. At first,

Fig. 8.33 Hidden animals

clearly varying yellow line thicknesses are noticeable. The thicker the line, the more yellow the overall image in this area appears, of course. Likewise, the image appears bluer with thinner yellow lines. Nicholas Wade has taken advantage of this hidden freedom of design of the line thicknesses in his work of art.

Due to the pure thickening and dilution of the yellow lines, he has incorporated a historical portrait of Ludimar Hermann (1838–1914), the discoverer of the grid shown here—and this is almost invisible to the observer at first glance and without impairing the flicker effect. If you do not see the portrait clearly, just look at the picture from a few meters away and squint your eyes.

In the right part of Fig. 8.34 a portrait of Karl Ewald Konstantin Hering (1834–1918) is hidden in the same way in a grid with red lines and green background. He was also significantly involved in the research of the Hermann grid. Here, too, the portrait was almost imperceptibly incorporated into the original image by varying the thickness of the red lines.

Before a well-deserved dinner, the bus stops at one last very special place: the legendary Scottish lake Loch Ness. Everyone talks in disarray when getting off and everyone knows a story about the loch's mysterious monster. Everybody is watching the calm water of the lake as shown in Fig. 8.35. Although no one really believes in the Loch Ness monster, everyone stays at a safe, respectful distance—you never know.

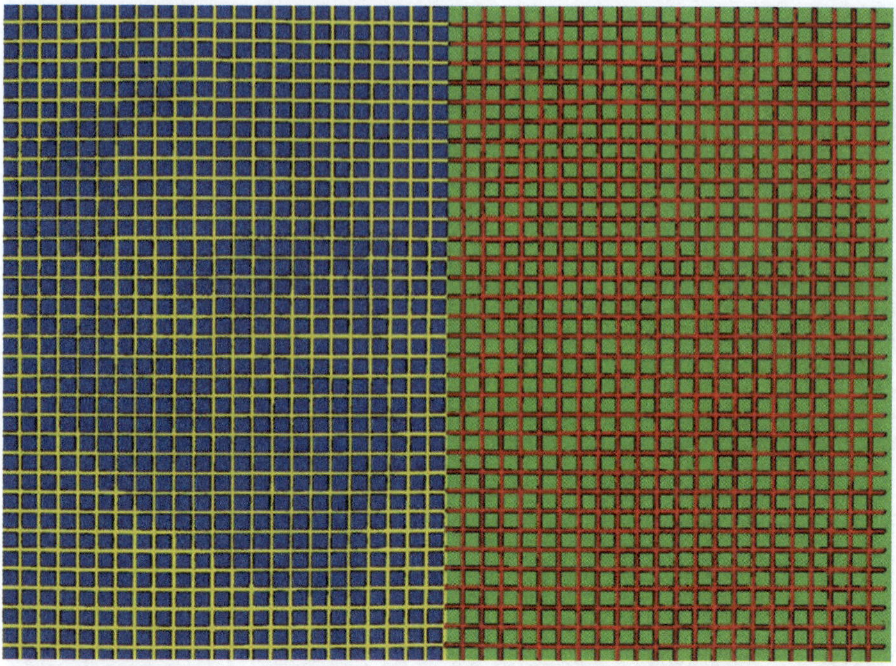

Fig. 8.34 Hermann's Lattice with hidden historical portraits of its discoverers Ludimar Hermann (left) and Karl Hering (right). The faces are easier to see if you look at the picture from a few meters away and pinch your eyes together. (Courtesy of Nicholas Wade)

And suddenly a fellow traveler shouts: "Actually, I see it now, it's really there, very clear and big!"

He describes to the others the location of the perfectly camouflaged animal:

"You just have to look a little deeper, through the surface of the lake. It's in the lower half and spread out all over the picture!" In time, more and more people actually see the beast in the deep.

The bus driver now helps the last doubters to get their bearings. He explains, "This picture is again an autostereogram, which we already met on the sixth journey (Sect. 6.7 and Fig. 6.35). In contrast to the autostereograms considered so far, this is a natural autostereogram with a natural pattern as texture, which repeats itself horizontally a few times periodically. If you look very closely at Fig. 8.35, you will see that the individual patterns repeat seven times with minimal variations!"

And it is exactly in these variations that the monster lies hidden—the perfect camouflage!

Everyone is now reassured that the whole thing is just a spatial illusion, and after a while the monster turns into a tremendous appetite.

So the travel group heads for the restaurant at the lake and is happy about the fantastic travel experiences through the not-at-all grey everyday life—and the restaurant about fantastic turnovers.

Fig. 8.35 The Loch Ness monster

Closing Words

In our eight journeys through human perception, you have seen that the human visual apparatus always functions according to one and the same basic principle: *striving for the optimum result with minimum effort*. By means of a multitude of optical illusions, ambiguous images, or spatial conflict situations, you could put this principle and your eyes to the test. Our perception has proven to be a master of compromise and always surprisingly imaginative. Thus, in some conflicts the simplest compromise solution was a temporal change of perception between the different possible alternatives. For some other images, however, the compromise consisted of an internal correction of the perceived image sizes such as edge lengths, brightness, colors, or spatial depth.

Like our environment, our perceptual system is also an extremely unstable entity. As you have seen in the fourth journey, even a tiny sensory perception can put the vast brain into either a huge crisis or high spirits. For example, the perception of a single number can cause major changes in people—just think of the reactions to the announcement of exam grades, the sales figures of this book, or the lottery numbers.

Our brain is not only able to perceive its environment, but can also give information to the environment. Surprisingly, this transmission of information happens completely differently than the reception: A major part of the communication with the environment takes place via language, i.e., the acoustic sound waves. Light and other electromagnetic waves play no role in this process.

But why then is there no output organ in humans for sending electromagnetic radiation? Bats, glowworms and swarms of beetles in Indonesia and elsewhere prove that it is in principle possible for nature to develop such a light-emitting organ. And the advantages over sound are obvious: Light is significantly faster and not as susceptible to disturbances such as headwinds. In addition, the fantastic human visual system can receive light even from several kilometers away.

One conceivable explanation for this omitted possibility of evolution is that language is a very new invention of nature. In the completed rapid development phase of language—and with it also of the brain—it was probably most expedient for nature in the "short" time available to fall back on at least rudimentary existing resources: namely the vocal cords and the tongue.

T. Ditzinger, *Illusions of Seeing*, https://doi.org/10.1007/978-3-030-63635-7

Over the last few centuries, the excellent development of the brain has put humans in the new, dangerous situation of being able to adapt their environment more and more to themselves. As a result, the balance of the world, which has been created over billions of years with the help of living beings of the most diverse species, is becoming increasingly unstable. It has become easier and easier for people to throw their environment out of balance, in extreme cases by pressing a single button! It is therefore of crucial importance to be aware of the significance of this balance, which is also life-sustaining for humans. Our world consists of an admirable harmonious interplay of the smallest things and events. Therefore, we should pay special attention to these smallest things and take their warning signs seriously in order to preserve our fantastic, fascinating, and wonderful world.

With this we have reached the end of our journeys into the land of perception. As is well known, a journey was only really good if you say at the end: "The homecoming is the best." Nevertheless, we hope that you will keep many new impressions and interesting memories from this wonderland of nature. It is my wish for you that even in the sometimes grey everyday life you will remember this fantastic colorful world of your own perception and that you will be able to enjoy it and yourself again and again!

At the end of this journey I would like to thank all those who were involved in the preparation and implementation or who suffered under it: Karl Zeile, Nicholas Wade, Tobias and Katrin Traber, Michael Stadler, Werner Skolaut, Ralf Schweller, Marion and Klaus Schreiner, Roland Schreiber, Dirk Reimann, Michael Rapp, Ken Quinn, Mariana Price, Baingio Pinna, Norbert Müller, William McLean, Ursi Maier, Rainer Lutz, Hajo Luers, Giuseppe Leonardi, Armin Kuhn, Akiyoshi Kitaoka, Scott Kelso, Petra Jantzen, Manuela and Jochen Holtz, Eva Hestermann-Beyerle, Fabian Herrmann, Franz-Josef Heimes, Sandra Heinzmann, Rainer Handel, Hermann Haken, Susanne Häfele, Martina Grupp, Thomas Gotthardt, Armin Fuchs, Rudolf Friedrich, Pat Foo, Jutta Förstner-Kuhn, Robert Fischer, Jacqueline Fernandes, Detlef Emeis, Robert and Irene Ditzinger, Winnie Devensky, Vince Billock, Michael Bestehorn and Junes, Mathis, Floris, Leonie, Yannic—and Sabine. A big thank you also goes to the publishing team Lilith Dorko, Marion Krämer, Katharina Neuser-von Oettingen, Anja Groth, and Ina Melzer for the good tour guide, various support, proofreading and the excellent cooperation, through which this book has gained a lot.

Suggestions for improvement, comments, criticism, praise, gifts, new optical illusions, or fantastic holiday postcards should be sent to the author: Thomas Ditzinger, Weingärtenstraße 19, 74934 Reichartshausen, Germany or electronically to tditzinger@web.de.

Picture Credits

All pictures and photos not listed here were newly created by the author himself.

Fig. 1.5: Dallenbach, American Journal of Psychology (1951).
Fig. 1.11: Ernst, B. (1989). Das verzauberte Auge. Köln: Taschen Verlag.
Fig. 2.3: Gillam, B., Geometrisch-optische Täuschungen. In: Wahrnehmung und Visuelles System, Spektrum-der-Wissenschaft-Verlag. Heidelberg, 1986.
Fig. 2.11: Gillam, B., Geometrisch-optische Täuschungen. In: Wahrnehmung und Visuelles System, Spektrum-der-Wissenschaft-Verlag. Heidelberg, 1986.
Fig. 2.19: Schober, H., Rentschler, I. (1972). Das Bild als Schein der Wirklichkeit. Optische Täuschungen in Wissenschaft und Kunst. München: Moos.
Fig. 2.28: Akiyoshi Kitaoka: Cushion (1998).
Fig. 2.29: Akiyoshi Kitaoka: The eyes (2002).
Fig. 3.14: US Postal Service (1996).
Fig. 3.20: Akiyoshi Kitaoka: The music (2002).
Fig. 3.27: nach Adelson (1993). Perceptual organization and the judgement of brightness. Science 262, 2042–2044.
Fig. 3.28: White M. (1981). The effect of the nature of the surround on the perceived lightness of gray bars within square-wave test gratings. Perception 10: 215–230.
Fig. 4.1: unknown author, found in a toilet in a student dormitory in Freiburg.
Fig. 4.2: Rubin, E. (1921). Visuell wahrgenommene Figuren. Kopenhagen: Gyldendalske.
Fig. 4.3: left. Signet of the Chess World-Championship, 1990, right: Nader, S., In: Shephard, R., Mindsights. Freeman, New York, 1990.
Fig. 4.4: after Necker, L. (1832). Observations on some remarkable phenomenon which occurs on viewing a figure of a crystal or geometrical solid. The London and Edinburgh Philosophical Magazine and Journal of Science 3, 329–337.
Fig. 4.6: Kopfermann, H. (1930). Psychologische Untersuchungen über die Wirkung zweidimensionaler Darstellungen körperlicher Gebilde. Psychologische Forschung 13: 293–364.

T. Ditzinger, *Illusions of Seeing*, https://doi.org/10.1007/978-3-030-63635-7

Fig. 4.7: Staircase after Schröder, H., Über eine optische Inversion bei Betrachtung verkehrter physikalischer Bilder. Poggendorfs Annalen der Physik und Chemie 181 (1858) 298–311. Wundt'scher Ring und Mach'sches Buch in: Gräser, H., Spontane Reversionsprozesse in der Figuralwahrnehmung. Dissertation. Trier 1977.

Fig. 4.9: Stadler, M., Kruse, P. (1995). The function of meaning in cognitive order formation. In: Kruse P., Stadler M., Ambiguity in mind and nature, 5–21.

Fig. 4.11: Coren, S., Ward, L. (1989). Sensation and perception, 3rd edition. San Diego: Harcort, Brace, Jovanovich.

Fig. 4.15: Botwinick, J. (1959). Husband and father—A reversible figure. American Journal of Psychology 74, 321–313.

Fig. 4.17: Hill, W., My wife and my mother in law. Puck, 6. November 1915.

Fig. 4.19: Stadler, M.

Fig. 4.20: Delfin oder Känguru? Nach einer Idee von John F., Kihlstrom.

Fig. 4.21: Jastrow, J. (1900). Fact and fable in psychology. New York: Houghton Mifflin.

Fig. 4.22: Stadler M., Kruse, P. (1995). The function of meaning in cognitive order formation. In: Kruse, P., Stadler, M., Ambiguity in mind and nature, 5–21.

Fig. 4.23 left: Verbeek G., In: Sunday, New York, Herald, 1900.

Fig. 4.24: "Squirrel or Swan?", und "Seal or Donkey?" from: Fisher, G. (1968). Ambiguity of form: Old and new. Perception and Psychophysics. 4/3, 189–192. "Rat or Man?" after Bugelski, B., Alampay, D. (1961). The role of frequency in developing perceptual sets. Canadian Journal of Psychology 15, 205–211.

Fig. 4.25: Schreiber, R.

Fig. 4.27: Fisher, G. (1967). Measuring ambiguity. American Journal of Psychology 80, 541–547.

Fig. 4.28: Giuseppe Arcimboldo, 1563.

Fig. 4.29: Robert Fischer: Ein neuer Tag, 1991/1992.

Fig. 4.30: Wolfram Nagel: Katze und Vogel, 1997, Löwe und Gesicht, 1997.

Fig. 4.31: Silke Haarer: Metamorphose, 1994.

Fig. 4.32: Kuniyoshi. A Face of Bodies, 1990.

Fig. 5.12: Gregory, R. (1972). Auge und Gehirn. Frankfurt: Fischer-Verlag.

Fig. 5.17: Kanehara Shuppan, Co. Ltd.

Fig. 5.25: Vic Winter, ICSTARS Astronomy Inc., 2001.

Fig. 5.29: Akiyoshi Kitaoka: Green and Blue Spirals, 2003.

Fig. 5.32, Fig. 5.33, Fig. 5.34: from Pinna, B. (1987). Un effetto di colorazione, in: Il laboratorio e la città. XXI Congresso degli Psicologi Italiani, Majer, V., Maeran, M., and Santinello, M., 158. Pinna, B., Brelstaff, G., and Spillmann, L. (2001). Surface color from boundaries: A new 'watercolor' illusion, Vision Research 41, 2669–2676. Pinna, B., Werner, J. S. and Spillmann, L. (2003). The watercolor effect: A new principle of grouping and figure-ground organization, Vision Research 43, 43–52.

Fig. 6.23: Rainer Handel.

Fig. 6.32: Franz-Josef Heimes, 2004/2005.

Fig. 6.35: AutoVision (Ditzinger, T., & Kuhn, A.). (1994). *Phantastische Bilder. Image-3D*. München: BHV Verlag.
Fig. 6.37: Dirk Reimann.
Fig. 6.42: with kind permission of Carl Zeiss, Oberkochen. taken from: Schöppe G., Danz, R. (1993). Das Jenamap—ein Mikroskopsystem für automatische Phasenmessungen. Zeiss Informationen 3, 18.
Fig. 7.5: after John Ayrton Paris, 1826.
Fig. 7.7: MacKay, D. M. (1957). Moving visual images produced by regular stationary patterns. Nature 180, 849–850.
Fig. 7.8: Akiyoshi Kitaoka: Warp, 2003.
Fig. 7.9: Wade N, (1982) The art and science of visual illusions, Routledge, Kegan & Paul, verbesserte Version Wade N, (1990) Visual allusions: Pictures of perception, Psychology Press Lawrence Erlbaum.
Fig. 7.14: Musatti, C. L. (1924). Sui fenomeni stereocinetici Archivo Italiano di Psicologia 3, 105–120.
Fig. 7.15: Ouchi, H. (1977). Japanese optical and geometrical art. Mineola, NY: Dover.
Fig. 7.16: after Akiyoshi Kitaoka: Out of Focus, 2001.
Figs. 7.17 and 7.18: Pinna, B., Brelstaff, G. J. (2000). A new visual illusion of relative motion. Vision Research 40, 2091–2096.
Fig. 7.19: Akiyoshi Kitaoka: Spa, 2003.
Fig. 7.20: Akiyoshi Kitaoka: Apples 2, 2004.
Fig. 7.21: Akiyoshi Kitaoka: Two Rings, 2005.
Fig. 7.22: Akiyoshi Kitaoka: Heat Devil, 1998.
Fig. 8.12: based on a template from Beau Lotto.
Fig. 8.16: Bayer 04 Leverkusen Fußball GmbH.
Fig. 8.22: Petra Jantzen.
Fig. 8.23: Edward Adelson (1995).
Fig. 8.24: Beau Lotto.
Fig. 8.25: Beau Lotto.
Fig. 8.32: Martin Liebetruth, SUB Göttingen.
Fig. 8.34: Nicholas Wade (2011).
Fig. 8.35: AutoVision (Ditzinger, T., & Kuhn, A.). (1994). Phantastische Bilder. Image-3D. München: BHV Verlag.

References[1]

[1]The following eight books are particularly recommendable and useful for various reasons and cover the seven journeys in breadth thematically well. You will then find a selection of specialized books and articles on the individual chapters.

Metzger, W. (2009). *Laws of seeing*. Cambridge: MIT Press. Classic, excellent, and long out of print. I was once told about the original German edition of this book, from a competent source, "If you see this book in a library, be sure to steal it." Fortunately, this is no longer necessary.

Falk, D., Brill, D., & Stork, D. (1990). *Seeing the light: Optics in nature*. Basel: Birkhäuser Springer. Excellent book on light and vision from the perspective of physics.

Goldstein, E. B. (2007). *Cognitive psychology: Connecting mind, research and everyday experience* (2nd ed.). Belmont: Wadsworth Inc Fulfillment. Outstanding textbook of perceptual psychology for students and lecturers.

Gregory, R. (1997). *Eye and brain: The psychology of seeing* (5th ed.). Oxford: Oxford University Press. Competent presentation of visual processes and illusions from the perspective of a well-known brain researcher with many original works.

Robinson, J. O. (1998). *The psychology of visual illusion*. Mineola: Dover. Original, valuable observations of optical illusions from the perspective of psychology.

Ninio, J. (2001). *The science of illusions*. Ithaca: Cornell University Press. Exciting and competent presentation of optical illusions in text and pictures.

Frisby, J. (1979). *Seeing: Illusion, brain, and mind*. Oxford: Oxford University Press. A well-founded ramble through the optical illusions and their perception.

Block, J. R., & Yuker, H. E. (1988). *Can you believe your eyes?: Over 250 illusions and other visual oddities*. New York: Taylor & Francis Ltd. Entertaining, instructive, and well-structured book with short explanations for each of the 250 deceptions presented.

First Journey: Light, Perception and the Laws of Seeing

Kandel, E., Schwartz, J., & Jessell, T. (2000). *Principles of neural science* (4th ed). New York: McGraw Hill.

Koffka, K. (1999). *Principles of Gestalt psychology* (new edition). Oxford: Routledge.

Köhler, W. (1992). *Gestalt psychology* (new edition). New York: Liveright.

Wertheimer, M. (1912). Experimentelle Studien über das Sehen von Bewegung. *Zeitschrift für Psychologie, 61*, 161–265.

T. Ditzinger, *Illusions of Seeing*, https://doi.org/10.1007/978-3-030-63635-7

Second Journey: The Geometric-Optical Illusions

Fraser, J. (1908). A new visual illusion of direction. *British Journal of Psychology, 2*, 307–320.

Gillam, B. (1986). Geometrisch-optische Täuschungen. In *Wahrnehmung und Visuelles system* (pp. 48–57). Heidelberg: Spektrum-der-Wissenschaft-Verlag.

Kaufman, L., & Rock, I. (1962a). The moon illusion I. *Science, 136*, 953–961.

Kaufman, L., & Rock, I. (1962b). The moon illusion. *Scientific American, 207*, 120–132.

Rock, I., & Kaufman, L. (1962). The moon illusion II. *Science, 136*, 1023–1031.

Third Journey: Perception of Forms and Brightness

Adelson, E. (1993). Perceptual organization and the judgement of brightness. *Science, 262*, 2042–2044.

Craik, K. (1966). *Nature of psychology*. Cambridge: Cambridge University Press.

Cornsweet, T. (1970). *Visual perception*. New York: Academic Press.

Hermann, L. (1870). Eine Erscheinung simultanen Contrastes. *Pflügers Archiv für die gesamte Physiologie, 3*, 13–15.

Julesz, B. (1986). Texturwahrnehmung. In *Wahrnehmung und Visuelles System* (pp. 48–57). Heidelberg: Spektrum-der-Wissenschaft-Verlag.

Julesz, B. (1971). *Foundations of cyclopean perception*. Chicago: University of Chicago Press.

Kanizsa, G. (1955). Marzini quasi-percettivi in campi con stimolozione omogenea. *Rivista di Psicologia, 49*, 7–30.

Mach, E. (1914). *The analysis of sensations*, republished 1999. Bristol: Thoemmes Continuum.

O'Brien, V. (1959). Contrast by contour enhancement. *American Journal of Psychology, 72*, 299–300.

White, M. (1979). A new effect of pattern on perceived lightness. *Perception, 8*(4), 413–416.

Fourth Journey: Ambiguous Perceptions

Botwinick, J. (1959). Husband and father-in-law: A reversible figure. *American Journal of Psychology, 74*, 312–313.

Bugelski, B., & Alampay, D. (1961). The role of the frequency in developing perceptual sets. *Canadian Journal of Psychology, 15*, 205–211.

Fisher, G. (1967). Measuring ambiguity. *American Journal of Psychology, 80*, 541–547.

Fisher, G. (1968). Ambiguity of form: Old and new. *Perception and Psychophysics, 4*(3), 189–192.

Haken, H. (1977). *Synergetics*. Berlin: Springer Verlag.

Hartmann, H., & Heiß, R. (1962). Zur psychologischen Bedeutsamkeit der optischen Inversion. *Diagnostica, 8*, 23–38.

Hill, W. (1915, November 6). My wife and my mother-in-law. Puck.

Jastrow, J. (1900). *Fact and fable in psychology*. New York: Houghton Mifflin.

Kruse, P., & Stadler, M. (1995). *Multistability in cognition*. Berlin: Springer Verlag.

Künnapas, T. (1957). Experiments on figural dominance. *Journal of Experimental Psychology, 53*, 31–39.

Leeper, R. (1935). A study of a neglected portion of the field of learning—The development of sensory organization. *Journal of Genetic Psychology, 46*, 41–75.

Necker, L. A. (1832). Observations on some remarkable phenomenon which occurs on viewing a figure of a crystal or geometrical solid. *The London and Edinburgh Philosophical Magazine and Journal of Science, 3*, 329–337.
Oyama, T. (1960). Figure-ground dominance as a function of sector angle, brightness, hue and orientation. *Journal of Experimental Psychology, 60*, 299–305.
Pöppel, E. (2001). *Lust und Schmerz* (Neuauflage). München: Goldmann Verlag.
Rubin, E. (1921). *Visuell wahrgenommene Figuren*. Kopenhagen: Gyldendalske.
Verbeek, G. (1900). *Sunday*. New York: Herald.

Fifth Journey: The Colors and the Grey Everyday Life

Benham, C. E. (1894). The artificial spectrum top. *Nature, 51*, 113–114, 200.
Bidwell, S. (1899). *Curiosities of light and sight*. London: Swan, Sonnenschein and Co.
Brown, P., & Wald, G. (1964). Visual pigments in single rods and cones of the human retina. *Science, 144*, 45.
Ditzinger, T., Billock, V., Holtz, J., & Kelso, J. A. S. (2000). The leaning tower of Pisa effect. *Perception, 29*(10), 1269–1272.
Helmholtz, H. (1911). *Handbuch der physiologischen Optik*. Hamburg: Voss Verlag.
Hering, E. (1878). *Zur Lehre vom Lichtsinne*. Wien: Gerold Verlag.
Ishihara, S. (1959). *Test for colour-blindness*. Tokyo: Kanehara.
Marks, W., Dobelle, W., & MacNichol, E. (1964). Visual pigments in single primate cones. *Science, 143*, 1181.
Nathans, J., Thomas, D., & Hogness, D. S. (1986). Molecular genetics of human color vision: The genes encoding blue, green and red pigments. *Science, 232*, 193–202.
Pinna, B. (1987). Un effetto di colorazione. In V. Majer, M. Maeran, & M. Santinello (Eds.), *Il laboratorio e la città. XXI Congresso degli Psicologi Italiani* (p. 158).
Pinna, B., Brelstaff, G., & Spillmann, L. (2001). Surface color from boundaries: A new 'watercolor' illusion. *Vision Research, 41*, 2669–2676.
Pinna, B., Werner, J. S., & Spillmann, L. (2003). The watercolor effect: A new principle of grouping and figure-ground organization. *Vision Research, 43*, 43–52.
Pulfrich, C. (1922). Die Stereoscopie im Dienste der isochromen und hetero-chromen Photometrie. *Naturwissenschaft, 10*, 553–564.
Purkinje, J. (1825). *Neuere Beiträge zur Kenntnis des Sehens in Subjectiver Hinsicht*. Berlin: Reimer.
Vernes, J. (1882). *Le Rayon vert (the green flash)*. L. G. F.
von Goethe, J. W. (1810). *Zur Farbenlehre. Erster Band, Abschnitt 52*. Tübingen: Cotta. new edition 2003, Verlag Freies Geistesleben.

Sixth Journey: Spatial Vision

Julesz, B. (1971). *Foundations of cyclopean perception*. Chicago: University of Chicago Press.
Pulfrich, C. (1922). Die Stereoscopie im Dienste der isochromen und hetero-chromen Photometrie. *Naturwissenschaft, 10*, 553–564.
Reimann, D., Ditzinger, T., Fischer, E., & Haken, H. (1995). Vergence of eye movements and multivalent perception of autostereograms. *Biological Cybernetics, 73*, 123–128.
Tyler, C., & Clarke, M. (1990). The autostereogramm. *SPIE Stereoscopic Display and Applications, 1258*, 182–196.

Seventh Journey: Movements Are Life

Aubert, H. (1887). Die Bewegungsempfindung. *Pflügers Archiv für die Gesamte Physiologie des Menschen und der Tiere, 40*, 459–479.

Ditzinger, T., Billock, V., Holtz, J., & Kelso, J. A. S. (2000). The leaning tower of Pisa effect. *Perception, 29*(10), 1269–1272.

Duchamp, M. (1935). *Rotoreliefs*.

Duncker, K. (1929). Über induzierte Bewegung. Ein Beitrag zur Theorie optisch wahrgenommener Bewegung. *Psychologische Forschung, 12*, 180–259.

Exner, S. (1887). Einige Beobachtungen über Bewegungsnachbilder. Centralblatt für Physiologie, 1.

Hubel, D. H., & Wiesel, T. N. (1959). Receptive fields of single neurons in the cats striate cortex. *Journal of Physiology, 128*, 574–591.

Kitaoka, A., & Ashida, H. (2003). Phenomenal characteristics of the peripheral drift illusion. *Vision, 15*, 261–262.

MacKay, D. M. (1957). Moving visual images produced by regular stationary patterns. Nature 180, 849–850.

Musatti, C. L. (1924). Sui fenomeni stereocinetici. *Archivo Italiano di Psicologia, 3*, 105–120.

Ouchi, H. (1977). *Japanese optical and geometrical art*. Mineola, NY: Dover.

Pinna, B., & Brelstaff, G. (2000). A new visual illusion of relative motion. *Vision Research, 40*, 2091–2096.

Pomerantz, J. (1983). The rubber pencil illusion. *Perception and Psychophysics, 33*, 365–368.

Rechtschaffen, A., & Mednick, S. A. (1955). The autokinetic word technique. *Journal of Abnormal and Social Psychology, 51*, 346, 298.

Reichardt, W. (1957). Autokorrelationsauswertung als Prinzip des Zentralnervensystems. *Zeitschrift für Naturforschung, 12b*, 447–457.

Schweizer, G. (1857). Über das Sternenschwanken. I. *Bull de la Societe Imperial des naturalistes, 30*, IV440–IV457.

van Santen, J. P. H., & Sperling, G. (1985). Elaborated Reichardt Detectors. *Journal of the Optical Society of America A, 2*, 300–321.

von Humboldt, A. (1851). *Kosmos III*. Stuttgart & Augsburg: Cottasche Buchhandlung.

Wade, N. (1982). *The art and science of visual illusions*. London: Routledge Kegan & Paul.

Wade, N. (1990). *Visual allusions: Pictures of perception*. London: Psychology Press, L. Erlbaum Associates.

Eighth Journey: Everyday Life Is Not Grey At All: Deceptions in Our Daily Life

Ditzinger, T. (2004a, October). *Farbkompetenz in der Praxis: Farbwahrnehmung und Zahnfarbnahme, dentallabor* (pp. 1581–1588). Planegg: Verlag Neuer Merkur.

Ditzinger, T. (2004b). Farbkompetenz in der Zahnfarbbestimmung. *Zahnärzt-liche Mitteilungen zm, 1,1.1*, 34–35.

Ditzinger, T. (2004c). *Leistungen und Fehlleistungen unserer Wahrnehmung, dentallabor* (pp. 1385–1392). Planegg: Verlag Neuer Merkur.

Ditzinger, T. (2004d). So wirkt die Farbe auf Sie und Ihre Patienten. *Zahnärztliche Mitteilungen zm, 1,1.1*, 28–33.

Heller, E. (2004). *Wie Farben wirken* (6. ed). rororo.

Kingdom, F. A., Yoonessi, A., & Gheorghiu, E. (2007). The leaning tower illusion: A new illusion of perspective. *Perception, 36*(3), 475–477.

Wade, N. (2011). Eyetricks. *i-Perception, 2*, 486–501.

Index

T. Ditzinger, *Illusions of Seeing*, https://doi.org/10.1007/978-3-030-63635-7

Printed by Wilco bv, the Netherlands